We Went A Loggin'

WE
WENT
A LOGGIN'

by

Esther Gibbs

NORTH STAR PRESS

St. Cloud, Minnesota

Editorial adaptation by Pernina Oliver Burke.

The Library of Congress Catalog Number: 74-81723
The International Standard Book Number: 0-87839-017-0

Printed in the U.S.A., by The Sentinel Printing Company, St. Cloud, Minnesota. This book was bound by The Midwest Editions, Minneapolis, Minnesota.

For further information address:
North Star Press
Post Office Box 451
St. Cloud, Minnesota, U.S.A.
56301

Contents

Prologue

Both women in the Little Partners Cafe were bone tired. Since Pearl Harbor every day was rough. A hundred thousand men had poured into the Great Lakes Naval Station next door — navy men and construction men, all taken up in a frenzy of building. The Little Partners Cafe had been opened so the executives of the surrounding industrial plants might have a place to eat and relax, but after Pearl Harbor there was no holding back the tide of hungry workers. Every Chicago eating place was jammed and the help worked to death. Esther and Eva sometimes wondered if their fantastically good setup, when they started, was worth it. It should have been a cafe with a little "class" — but that had long since vanished. At first that was the atmosphere, but the teachers watched their figures and their purses, and the businessmen with their ulcers were even worse. Esther complained that her inspiration to be a good cook was dying a slow death, and her bank account wasn't much better.

Then the defense workers hit town! They took over everything. The teachers retreated in confusion; the executives needed tranquilizers; the Little Partners were dead on their numb feet by noon every day — and had to go until night; but their bank accounts did prosper.

The early October evening was beautiful and calm, the work day done, as Eva snapped off the lights and said, "Let's get the books checked out, Esther, and get home. I'm dead." Esther's flat, tired voice answered, "Just tramp the money in a bushel basket. That's what

I told a customer we did, when he asked how long it takes to count our money these nights." She smiled wryly.

The door opened and a tall man stepped through. Eva cussed under her breath, "Forgot to lock the damn door!" Aloud, she said, "Sorry, sir, we are closed. We just forgot to lock the door." But he came right on as if he hadn't heard. He said, "Mrs. Jimmie?" He asked it like a question, but his voice said he knew.

Esther never looked up. She closed her eyes. She'd heard that one could meet everyone they knew, in the great human tide which engulfed wartime Chicago. Years ago, she'd been told, "You'll be blind before you're thirty" — so she'd learned to recognize and remember voices. To her they were like fingerprints — unique for each person. Now when the man said "Mrs. Jimmie," she knew where he fitted into her life. She looked up at him and said, "You were about to say, 'I'm the timber cruiser; I'll be here for supper,'" at which they both laughed.

"Well," he said, "I was here for lunch and told the waitress I knew you and would be back. I didn't mean to be this late."

The little cafe forgotten, they both were back in that logging camp in northern Wisconsin, twenty years before. Bob Merritt was cruiser for the lumber companies and showed up about once a month throughout the winter. He had heard of this girl in old Fred's camp who had talked her way into the camp cook's job. Women were bad luck in the woods, and crusty old Fred was the last man in the world to consider this kind of trouble.

Esther introduced the tall man. "Eva, this is Bob Merritt, from another world and another time. You get a cab, take all this home with you, and do the bookwork. We'll probably be right here when you come back in the morning. I feel a long visit coming on."

We were there, all right, the next morning. We had consumed gallons of coffee. The pies were baked for the day; we had eaten one for breakfast.

It was four in the morning. By five the early breakfast horde would be downing bacon and eggs. But what Bob had really wanted to know was how "Mrs. Jimmie" had come to be in that cut-over Wisconsin Chippewa Valley country.

"Well," I said, "I'll tell you the story and make it as short as I can."

"First of all, I want you to know I never worked except in my own home since camp days, until that heart attack took Jim. Bobbie, the little blonde boy I'd had with me in camp, had become a husky man, but he fell from a construction job right here at Great Lakes.

I didn't know what I could do to live, but I knew this was the time to find out. So when a friend asked, 'Esther, you can cook,' I found my new life. This little cafe became my new life and love. However, I do hope to *end* my days deep in the woods in a little log cabin, with the creatures of the forest as my pets — with no phones to ring, and where even God can't find me, and with enough books to occupy the rest of my life."

"Yes, Mrs. Jimmie, but you still haven't told me how *you* happened to be in that logging camp in the first place — or how you ever chanced to hit that burned-over, cut-over part of the country."

"Well, Bob, that was the question everyone asked everyone else. Each answer was different, but one thing we all agreed — we must have been crazy."

"Work was scarce after World War I. People were desperate to get out of the cities and onto some land where they could at least stay alive. Finding that land was the problem. The men did the scouting. It's a cinch no woman would have fallen for the salesman's pitch. Farm produce prices and land prices had been good all through the war years. Although the farmer was still considered a hayseed, the call of the land was strong in my man."

"The land in the old established farm areas was too high-priced for the average man to even consider, so when they saw ads in the Iowa town papers like this:

> Come to the beautiful Chippewa River Valley; Build yourself a home away from all the noise and worry of the city. We will sell you a 40 or 80 acre piece of land complete with house, barn, cow, chickens, tools, and grass seed for a small down payment.
>
> — many a city man bought blindly."

"They say, Bob, that by 1922, five hundred families had answered that call and nearly every forty had a settler. I don't know where they came from, or why, but I do know the stories of a few. My Jimmie was back from the war. Like many Iowa people who had been farm-renters, he now saw a chance to own his own place. We had the down-payment and that was about it. The lowest price for the land was thirty dollars an acre, and everything else was extra. At eight percent interest, it kept us humping just meeting the interest payment."

"You know how that Wisconsin land was. It was either stony or sandy, depending upon what the ice cap had done to it. In the early timber days it must have been beautiful, but what was left was a broken, burned, and sickening landscape."

"After the war the recession began. Corn prices fell until it was burned for fuel. Failures of the banks led to the great 30's depression, and with it the promoters of the Chippewa Valley colonization effort all went broke."

The cruiser still couldn't understand how I could marry a man who wanted to live like that.

"Well, Bob, I'll tell you. My dad came from a family of ten children, all born on a poor southern Illinois farm. Usually everyone lived their entire lives in the area where they were born. But my dad was born to wander. He and my mother were in the land rush into Oklahoma territory and staked a homestead."

"On the day I was two months old, my mother died. Dad took my six-year-old brother and me back to Illinois. For the remainder of his life, his itching feet led him all over this country and Canada. Opening up new country was just what he liked. Dad was a big, good-looking guy. Being an organizer and doer, he made out fine wherever he went. My beautiful stepmother was a fine woman, but she had no taste for the rugged life. My brother left home at 15, lied his age, joined the army in Canada, and after four years of hell was killed in France just before the 1918 Armistice. So I was always Dad's right-hand boy. That's what I tried to be, for Dad loved boys — girls he only tolerated. But when it came to education, he didn't forget I was a girl — he thought educating women was a waste of time."

"I often wished I'd known more about my mother, or about the early days in Oklahoma, but from Dad I learned only scattered bits and pieces, and I've never even seen my mother's handwriting."

"I think I needed identity as a woman — and solid, deep roots. Jim seemed to answer those needs. He needed roots, too. He was home from the war and eager to celebrate living; but the celebration had to be a continuing challenge — a hard pact between him and his God. He was a *good* man, you know — honest and hard-working, and more than eager to settle down. I needed him, and he needed me. That's all there was to it."

The cruiser, Bob Merritt, and "Mrs. Jimmie" never met and talked again, but before he left, Bob had said he hoped that one day I would write down all about my winter in the logging camp. When I said "Oh, I couldn't do *that*," my little partner piped up with, "No, it probably can't be done, but she'll just go along in her funny little way and do it."

We Went A Loggin'

Chapter One

WE WENT A LOGGIN'

So necessity is the mother of invention? True. But necessity is also the mother of adventure. When someone says that I've had a colorful, adventurous life, I doubt that it's much different than anyone else's. Seems to me that just living a normal life span is an adventure. How could it be otherwise? I've seen people, when confronted by an unexpected situation, just lie down in the road and let the traffic of life flow by on either side while they play "dead dog." They avoid adventure by the simple process of refusing to face it. Their dull voices chant, "You've had such an adventurous life. Nothing ever happens to me." And so help me, nothing ever will. They won't let it.

I don't attract adventure. I'm just receptive to it. I'm not more brave than anyone else — just maybe more foolhardy. I like the extra dash of a gamble now and then. Losing doesn't distress me at all, but the winner's pleasure still is sweet.

So we went to live in that part of Wisconsin that God seemed to have forgotten. Then it seemed the devil had made things still tougher by generously sprinkling every foot of the soil with rocks. This scene of dreary desolation was painted rosy by the glib tongues of super-salesmen, who reaped a harvest of the savings of tired people eager to believe the stories the salesmen told them. The truth would have been much harder to believe. Most of them were thrifty, middle-class

1

workers who had always eaten well of simple foods, dressed neatly and comfortably, enjoyed inexpensive recreation, saving a little each payday toward a retirement farm. Now the nest egg had hatched. They had their farms — complete with mortgage, of course. The city and its way of life seemed far away. No regular paydays now. Very little cash money to spend — but very little to spend it on.

They were a hardy lot blessed, fortunately, with a good sense of humor. They could joke over being "took in."

"They told me these rocks were just on the surface. 'Just pick off what you see and that's all'," but the truth was that the frost heaved up a hard new rock crop to greet you each spring. The Widow Butler had been here over thirty years and she said she had picked as many this last spring as she had the first spring.

The salesman had said, "This Mrs. Butler has lived here over thirty years." You thought that pointed to its being a good country, but now you knew that Widow Butler had been left with nine kids to raise and she couldn't raise enough cash to get out of the darned place with that brood.

Yes, they said you could just gather that old rotten pulpwood and get $5.00 a cord for it. Sure looked easy! But you found the paper mill didn't want that stuff. They wanted pretty good pulp logs, of which you didn't have any on your place. If you did have any cordwood, you needed it to keep from freezing in the paper-thin houses they built for you to live in while you faced the tough winters.

It was impossible to keep warm. The houses were set up on concrete block or stone foundations — two-by-fours set up for frames, divided to make two rooms downstairs and two rooms upstairs. To the frame on the outside was nailed a single layer of rough boards. Over this was a cheap roofing material, scarcely heavier than tarpaper. The downstairs was finished with the cheapest grade of paper wallboard which buckled and sagged as the green boards writhed in the drying process. The upstairs was bare of wallboard and had an open area all around where the roof joined the wall proper. The windows had no frames — just strips of thin wood around the edges to hold the sash in place. The green wood floors were impossible to sweep, and were so uneven as to ruin any floor covering brought from city homes that might be laid on them.

"Yes," the salesman told the new colonists, "There's plenty of work. Roads to build — pickles and truck garden crops to raise for incoming canning factories." There was very little road building, for the state had small interest in building highways through the valley. There was nothing there they really wanted anyone to see!

The cool nights they had said were so good for sleeping brought frost any month of the year. The canning companies would surely seek a more favorable place to raise a crop.

There were few cars and fewer radios or telephones. The doctor had a car, and the mailman. I think there was only one radio in an area of fifty miles, and that worked by earphones. A telephone costs only $12.00 a year. Who had that much? Only the general store could afford a telephone, and they would take messages. You would get your message when you came into town on Saturday, or when someone came along your way who'd been by the store, and gave you your information.

Chapter Two

IT'S TIME TO PAY

It was in this setting that I found myself at eighteen with my husband and small son. Bob had been born with a rupture. Not a dangerous condition, but with no money, nor a place to earn any, it became a problem that haunted my nights and days. Gathering the necessary funds for Bob's operation became one of the great adventures of my life. But when we entered into that first winter I wanted money, not adventure. My life was already rich in adventure. It was adventure enough to be Mrs. Jimmie and mother to little Bob. It was adventure to paint rosy dreams of "our farm" in the years to come.

The cards were stacked against the post-war settlers. The old folks didn't have a chance. We'd all been plucked clean, but at least Jim and I were young enough to dream and hope.

But doubts began to creep in. How could I even manage this little thing — this hernia repair — living as we were living? What if I had more children who would need food and shoes and maybe medical care? Early in our life in Wisconsin, I began to doubt the wisdom of fighting the barren, rocky land for the things we must have.

When I bought an apple for Bob, I'd divide it in four parts, and each day for four days he'd have a small surprise. Bob didn't know — but I did — that there were places where trees hung heavy with great red apples. One apple for four days! It was a hell of a life!

When I bought the apple, my neighbor, Mrs. Tilton, turned quickly around. "I'd sure like a pound of weiners," she said, "But I'd just better look the other way." As we rode along together, I told Mary Tilton that I'd get out of this rut somehow — I'd be darned if I'd look the other way all my life, because I didn't have fifteen cents for a pound of weiners. Mary said, "I would, too, if I were as young as you. But me and my Chris can't make another start." Their savings had gone into the farm. Chris was a fine toolmaker, but by now his place in the city had been taken by a younger man, and with three children to support, he couldn't go back to the city without money for a start.

Old Cappy and Mary Martin, who lived on our back road, told me the story of their return from the west and the wide open spaces. Cappy thought he needed 160 acres to, as he put it, "take a good breath in." He had enough money to get started and old "JC" was a generous money lender. They called him the "Settlers' Jesus." Cappy and Mary went into town, considered getting a couple of good cows. Got to get started soon, and they wanted to talk it over with "JC".

"Cappy," advised JC, "you just drive home a dozen of the best over there in that feed lot. Your credit's good and I'll O.K. your feed bill, too. You are one of my best men, Cappy. Ought to have at least four horses, too. Anything you want. Don't worry about nothing, Cappy. I'm back of you."

Cappy felt mighty prosperous when he saw his herd in the home pasture. He felt like a big-time farmer as he went out by lantern light early in the morning to start the milking. He got a nice cream check, too, but it seemed like he had to endorse it right over to JC at the feed store. That cut-over land produced no dairy rations.

"Keep pouring in the feed," JC said. "There will soon be a profit and you'll wipe out that debt in no time."

Things went along without much change. Whispers went about among neighbors that JC wasn't so generous any more. He rode about prodding, criticizing, putting on a little pressure.

The first of many failures saw Cappy, very early one morning, driving his herd back into town. The skinny critters he had bought on credit were fat and sleek now. Then Cappy took back his horses. They looked better, too. Everything at Cappy's farm looked better, except Cappy and Mary.

JC says "Cappy, it's time to pay." Cappy can't pay now. Soon, Cappy hopes, when he don't have to buy so much feed he can pay off his debt. But JC knows it won't be so.

"Cappy," says JC, "I'll give you ten acres all clear of debt, back on the Hogan road. I'll take your 160 and animals, and you'll be free and clear."

Cappy's face turns white. He looks at Mary and she's trembling. He looks at JC. He don't look so big-hearted right now. He isn't going to slap Cappy on the back and say, "What can I do for you, Cappy? Anything you want, Cappy. Speak right up." Cappy tries to speak right up, but JC's eyes are cold. "It's *that*, Cappy, or else."

"I'll take it," Cappy says, and turns to lead Mary out. Two very old and beaten people.

JC had a good buyer for Cappy's place. "And stir yourself, Cappy. I want you off in a week."

The whispers became louder. Who is next? "Today I heard JC told Billy to move off by next week." And so the stories went, and change came to our little valley.

Little Lonny Mason hiked across the blueberry marsh to ask me if I wanted a pussycat. "We're leaving here soon's it gets a little warmer," Lonny said, "But I've got two problems. Got to find a home for Midas, my good old pussycat, and someone to give my funnies to."

Sometimes city friends sent us the newspapers, magazine sections and the funnies. These were passed around and around the neighborhood. After they had made the rounds, Lonny collected them, and he explained that every couple months he read them over again. "Just like new funnies," Lonnie said.

I told him it seemed that any of his friends ought to be happy to fall heir to his funny-paper collection, but Lonnie said, "Mrs. Jimmie, most everyone over our way is checkin' out or expects to before long. JC gave Mom fifty dollars and told her to leave. Mom says we will start for Tomahawk, soon's it's warm enough for Linnie and I to go barefoot. Mom's uncle Ben said he can feed us if we can manage to get to his place."

Ever since Ted Mason had died and left Mrs. Mason and the twelve-year-old twins, everyone had felt Mrs. Mason should go someplace else. She was young, pretty and clever. But how? I knew she was doing just what I'd do. She'd walk out, with her boys. The fifty dollars would be for food and unavoidable expenses on the road.

Maybe they'd get a lift once in a while, after they got on the main highway.

The talk among the men at the general store, my Jim said, concerned "Old Man" Sacksaw's wife. Seems she was telling the new mailman how hard money was to come by. She said she'd do anything to make a dollar. "Anything?" he asked, and she repeated, "Anything." And when he looked in her eyes he knew she meant just that. He said he didn't think she could offer anything worth a dollar.

Jim said the idea made him sick at his stomach.

But I said, "Mrs. Sacksaw isn't old. Those twins of hers just started to go to school. The thing that makes me sick at *my* stomach is that I might look just like her when I reach her age."

Chapter Three

DEVIL'S HOLE

That night Jim and I talked far into the night about our problems and our prospects. He loved the farm. So did I — That is, a real farm in real farming country. I pointed out that we only had one life to live in this glorious modern world, and I didn't want to miss all the better things of life. Too many of the good things, only money could give us. Not luxuries — just necessary things. It would even take money to turn this wild land into some semblance of a farm. Those great old pine stumps would endure for longer than we, and time would effect small change in the rocks that littered our future fields. Only dynamite would lift the stumps and rocks, and only money would buy that. That much money was impossible to find here.

Jim and I had both seen a lot of the world. We knew it could be otherwise. But my Jim was an optimist. Everything would be all right. I was to hear him say that many a time. I was also to learn that sayng it did not make it so.

Jim agreed we must have money, but he would get that as soon as the crops were planted. A little acreage here and there. Then Jim would be on his way. He would go as a hired hand to the western harvest fields, follow the crops, and send the money home. But as our hoard grew, so grew our taxes. If you fenced your forty, they thought you were rich, and the taxes went up. You couldn't win. There was no margin for gain.

Most of the neighborhood men were much worse off than Jim. If they went to work in the city, they had to take their families with them. It took all they earned to keep their family in the city. One winter under that plan, and they just stayed on the land and starved it out together. But Jim was free to go wherever a dollar could be earned. Cards came to our rural mailbox from the Dakota harvest fields, a pea cannery in Southern Wisconsin, the beet fields of western Minnesota. Jim travelled where there was cash work to do, and he never paid a fare. But it was dangerous riding freights.

I worried a lot. But I kept busy — too busy to really worry myself sick. I gathered pine knots for fuel, filling a gunnysack full and dragging it behind me to the house. I emptied sackful after sackful in a shed near the house and back to the woods for more. The more knots I gathered, the further I had to go to fill my gunnysack. I always had my rifle and shells along and any foolish game bird or animal that chanced along was my prey, in season or out. I canned game and anything else that was edible, including a wild green known as pigweed. I hated all greens, but they mght be vital to our health before the spring brought new green life.

I made fuzzy rugs from old gunnysacks — drapes, too. We learned to do this at the Farmers' Club, our only social function in that lonely countryside.

With the pine knot crop cleaned up once and forever, I started to cut the young birch and poplar that had sprung up. The young trees were perhaps twenty feet high as big around as my arm. I would cut two trees and put one under each arm. Then I walked to the house like a horse between buggy-shafts. At the shed I cut them up into stove lengths with a buck-saw.

I roamed our land hopefully, trying to get a shot at some stray deer. The year after Bobby was born, he rode in an old army sack on my back, a lunch for us rode in one pocket, shells in another. We two would wander all day, but I knew our chances were small. Even the animals seemed to avoid this territory. There was little food and really no shelter for the wild things, now that the great forests had been destroyed.

One warm, late spring day when Bobby was napping in his crib, I felt restless. But not wanting to get beyond hearing, I wandered to a spot that had always fascinated me. Local natives called it "Devil's Hole." It was a deep depression in the earth, perhaps twenty-five feet deep and seventy-five feet across, mostly bare of vegetation —

mostly, I guess, because it was lined with solid boulders, as if the devil had laid them in by hand. The forty acres of which this hole was a part, had been sold many times. Each new owner swore he had walked over every foot of that forty and that the cursed hole had not been there when he bought the land. Some had said, when they discovered the hole, that they had been tricked — that this was not the forty they had looked at; but every other landmark said it was.

The truth was that the salesman knew this land so well that he could take you over it a dozen times from every angle and never once reveal the "Devil's Hole." In fact, we who knew it was there, as I did, often had trouble coming upon it, so that sometimes we swore it was not really always there. Some neighbors said that when the hole was there the devil was abroad, and now there'd be the devil to pay; and when they couldn't find it they said the devil had gone and pulled his hole in after him.

Some of the old-timers told tall tales about the Devil's Hole to entertain, and to put a little scare into the poor timid greenhorns. A soft mist usually covered the hole from early evening, and many a morning sun shone on the results of fierce battles between rival bucks, often with their horns locked together, both dead, with their necks broken by the struggle.

Old John White claimed to have witnessed a fight to the death between two brown bears over the affection of a young female at the site of the Devil's Hole. Most of the local hunters shunned the area from a sense of fear of the unknown.

Being game warden used to be an unhealthy occupation for a man in this country. Some wardens simply disappeared while on duty. It was whispered that the Devil's Hole held the key to some of these disappearances.

The day I visited the Devil's Hole I took my rifle — which I carried on all inspection trips that went beyond the barnyard. As I neared the hole, I inched along on my belly until I could peer into the depression through some hazelbrush growing along its edge. I tried to make believe I could see a great stag fight, or more romantic, the mating of brown bears. Suddenly antlers seemed to take form. I thought my mind was playing tricks on me. Then the antlers seemed to move! There was only one way to find out if I was dreaming. With the rifle I drew a bead on a spot behind the antlers which would cut the spinal cord. I pressed the trigger and my shock was as great as that of the four-point buck that jumped to his feet and dashed blindly up the steep wall of the hole. After several attempts to escape, he collapsed as if dead. By this time I knew I wasn't dreaming. I raced wildly through the brush calling to my neighbor, old John Corcoran. When I saw him I panted, "John, I think I killed a deer in the Devil's Hole. I've got to cut his throat! Hurry! Bring a rope. And ask Nelly to come along and take care of Bobby!" I ran on home for Jim's hunting knife and to scoop up little Bobby, who was awake and fretting by this time. John soon joined me with his length of rope, and Nelly took over Bobby.

I could trust John to lower me into that hole with the rope around me, but I was afraid to cut the buck's throat. I had heard that those razorsharp hooves lashed out in the death convulsion. I couldn't hesitate too long, so I made a swift slash and jumped back, but it was well dead. Next I put the rope around the antlers and John pulled the buck up, with me struggling and pushing along behind. I sat down and trembled like a craven coward, and said, "Gee, John — the poor thing," and started to cry. John could only say his pet by-word, over and over again, "By gum! By gum! You're the dangdest kid I ever did know." I had to tell John all about the shot and surprising the deer.

While John finished cutting the animal's throat, to bleed it properly, I hiked off to get help. The two Bane brothers, whose farm adjoined ours, came quickly to help us with our deer. Soon the buck was hung and cleaned. The Bane's wives, who were sisters, and all the young Banes, came over to see the deer. They were amazed, and being city girls, they couldn't believe it — a girl had shot the deer! Shorty and Bud Bane guessed the weight of the buck at two hundred pounds.

John Corcoran went to work with his skinning knife, and when the fresh meat was revealed, the Banes families wanted me to treat them to venison steaks right then and there. Old John and I knew this wouldn't be a good idea, but we couldn't make them understand. Finally I said we'd split the deer four ways. John and I would get rear quarters and each of the Banes a front, and I would treat the whole gang to a feed from mine as fast as a big fire could be made. So we all set to the job.

It wasn't long before the Banes, long away from fresh meat, were sure putting it away. Everyone was gay. It wasn't deer season, but no one worried about that. The gods were good to us — and who were we to question a gift from the gods?

Shorty disappeared in the brush to reappear with a big paper star pinned on his chest. He came up, bit into a steak, and looking at me with his fierce blue eyes, said, "Is this venison you got for dinner, ma'am?" I gave the old backwoods answer: "Yes, and by God we'll have game warden for supper if you say anything." Everyone roared at our joke. All was gay "on the spur" that day. The four of us lived off the main road on a road called "the spur" by the main-roaders. They were a bit jealous of us because we all got along well together, while they seemed to have constant squabbles among themselves.

We had very little to do with the main-roaders. Their petty quarrels bored us. They seldom ventured onto our spur. Our road began at the bottom of a steep hill. Coming in, the spur was bad enough, but going out, climbing up that sheer hill, set both man and beast a-tremble. Cars never tried it. We could have cut down the hill, but we never got around to it. We really craved no company. Yes, sometimes the hill was tough on us, too, but we got together on getting loads up or down. We liked our privacy and kept our hill just as it was.

So we ate our venison with small fear of being hosts to unwelcome guests.

The Banes were finally satisfied and prepared to go home. Bud said, "I'll sleep good tonight. My innards is sure content." Old John and I smiled to each other as the two families went down the road.

"Sure will be a show at the Banes' when they all get traveling; going to strain their outdoor plumbing, but you might as well have tried to hold back a bunch of hungry lions." John chuckled and shook his head.

Next morning I just had to wander over to see how the Banes survived their ordeal. They were all pale, wan, and sort of wobble-geared. Bud Banes' wife wanted me to take back their quarter of venison. That didn't make me mad at all, and the next day I was busy cold-packing the venison, putting it under the house where Jim had made a dugout cellar for food storage.

The deer episode was a big thrill for me and all our neighbors, but being out of season, we had to keep still about the whole thing. Except I wrote of my hunting adventure to my Jim.

Jim's faith in our future burned ever bright. He wrote back, "With a wife like you, honey, I can't lose. You're the best little pioneer I ever saw."

"Yessir," wrote Jim, "we'd make a home on our northern Wisconsin land; he'd live to be a hundred years old, and then be buried under the great granite boulder out in the white birch grove."

We were young and we were in love. At least Jim was all of the time, and I was some of the time. At other times my violent temper boiled over. I guess it was too big a jump from being a little girl to being a wife. It frightened me. I wanted to lash out at everything, and when I couldn't vent my wrath on the fierce cold, the cutting winds, the scarcity of food, or the ugliness of the country, I lashed out in my fury at Jim.

Jim thought I was cute when I was mad, but my tongue was cruel and sharp and I would throw insults at him until I could see I had inflicted a severe hurt. Then I was sorry. Really sorry. I tried to make amends. Jim said, "I never knew a sweeter little devil." He forgave my anger easily. I think he should have slapped me down about then. I might have grown up quicker and saved us both a lot of misery. But that wasn't Jim's way. When I'd say, "Jim, I'm so sorry I was mean; when I get mad I just go loco," Jim would reply "Were you mad? Well, I didn't know that. Honest I didn't." Then he'd smile at me and his brown eyes would shine as he told me his plans for "next year." I was to learn that with Jim everything was going to be wonderful "next year."

Chapter Four

A STITCH IN TIME DON'T HELP

As Jim tried to make a farm, I tried to make a home.

I loved to cook, and was good at it. I liked to keep my house clean and make it pretty, and to be clean and neat myself. But I couldn't do any of these things well if Jim watched me, and that seemed to be his chief delight. My aunt, who taught me home-making skills, could cook and bake all day and be as neat as a pin, but I always had flour in my hair, wiped my hands on my apron, and turned into a general mess. I thought he was laughing at my clumsiness, and it upset me very much.

Cosmetics were completely new to me. I had never been allowed to use any at home, or even to put my hair up. I sent everywhere for free samples of powder and rouge (when I had money enough for penny postcards for the orders), and spent hours combing my hair and painting my face. Painting, I guess, is the right word. Jim always pretended I really looked beautiful, but in subtle little ways he tried to tone down my paint jobs. He had a beautiful sister, Katie, and he'd say, "Katie don't use rouge and I don't think you need any, either. You're pretty just natural," but I wanted to startle 'em and I guess I did.

But at one chore I was a total flop — that was sewing. And not for lack of opportunity. My aunt was a skilled seamstress. She had even made the suit my father wore at his wedding. She was patient and kind in her teaching, but she realized it was very difficult for me — made me nervous to the point of tears. I had told myself the simpler points of the art must be mastered before I should be a wife, but alas, I put it off and here I was — not only a wife but soon to be a mother.

The first winter I made quilts of anything available; all by hand, of course, in my big clumsy stitches. We had no sewing machine and I didn't know how to operate one, anyway. Aunt Lu had said I must learn the rudiments of sewing by hand before I could use the machine. These basic skills, I had never mastered to her satisfaction.

Jim, who had spent some time in the tailor shop at the Indian school, had far greater skill than I with the needle. Every evening found us working on quilt tops or curtains made from remnants. At last we had a flour-sack bedspread. I tried harder than ever to improve my sewing. I tried to make smaller stitches and to back-stitch to help make the finished piece stronger. It was embarrassing to have your smiling, brown-eyed husband do so much better than yourself, but Jim was generous with praise and encouragement.

I needed warm flannel pajamas. Buying ready-mades was out of the question. Jim told me to get some flannel and to make them but I said I couldn't do that. Finally I did get some flannel and Jim said, "Now you make one pair and I'll make the other and we will see which pair lasts longer." So it became a game for us. One day when he'd gone to town I finished mine — complete with pink chambray cuffs on pants and sleeves. I opened the door to greet Jim as if I were attired in the latest dinner dress. He put his arms around me and then noticed the strange softness of the new flannel. He feigned amazement, and his praise thrilled me so that I went to work the next day on the other pair, and trimmed that pair in blue.

The approaching birth of our son Bob — yes, Jim always knew it would be "Bob" — brought me face to face with a big problem — clothes. Those small, pretty things every mother proudly shows her friends. I just could not produce these things. I had thought that perhaps as motherhood came upon me, those skills, too, came naturally, but no such gentle miracle came to me. I cried because I was so clumsy. I had great need of home-making of every kind, and here I was completely lacking in a basic one.

Jim suggested we get started on Bob's clothes, and I turned frantic, pleading eyes on him. "Oh, Jim, I can't make those pretty little things like babies wear." But, somehow, things always work out, the Lord watching over fools as he does. A couple of years before, I had made a friend of a strange, quiet, Russian girl. She had married a well-to-do farmer, and, because she had known hunger and want long ago in her native land, she tried to help anyone she could, now that fate had been generous with her.

She invited me to visit her for a week. We were the same age — just in our late teens; but she might have been a hundred years old, she was so somber and quiet. I have often wondered if the week was as hard on her as it was on me. Her husband was nice, much more warm and human than she. His parents had adopted the tiny Russian girl, nursed her back to health and quiet young womanhood. They were proud that their son had married her and that they took over the operation of the family farm. Her home was lovely, but so dull! Everything was always in order, and you must not touch this or that! I remember that she loved children but had despaired of ever having any. When I saw their bedroom, decorated lovingly in pink and white, but with her bed at one end of the long narrow room and her husband's far to the other end, I remarked, "No wonder you are not having a baby. This distance is discouraging." Her face burned red, but I think her husband favored the suggestion.

Anyway, this kind-hearted girl bought material and made a complete layette for our baby. Everything and anything, all new and all finished except the buttonholes. She thought that I'd like to make them!

So my big problem was solved quickly by this shy Russian girl.

They brought me home, along with the new little clothes and other things for myself and Jim, and even some small rugs for our little house.

When I told Jim I'd sew snaps on the baby-clothes, because I couldn't make buttonholes, he demonstrated that he knew how to make buttonholes, and sat until two-thirty one morning, finishing them all. I felt ashamed for being such a helpless wife, but at least I tried to stay awake and keep Jim company and keep fire in the stove while he sewed.

Chapter Five

SWAMP RATS EAT WELL

But, if my fingers were all thumbs, and my stitches a bit too big, I more than made up for this lack in my other chores. With my love of cooking, I had a knack of making a meal from almost nothing, which often came in handy — nothing often being the main ingredient. I started from nothing and tried to produce something.

Short season root crops, potatoes, onions, and rutabagas were the mainstay of the settlers throughout the long winter. One neighbor described his diet as potatoes and onions one day, and onions and potatoes the next. Maybe someone had only rutabagas, he'd trade you some "beggies" for potatoes and onions. How good to have just a little variety in our diets! The old-timers ("swamp rats" we called them) seemed to fare best gathering food supplies. They knew the country better than we did; knew where to find a few berries, or, perhaps, some wild honey. This woods skill was completely beyond the greenhorn "City Jake." Besides, he'd get lost in some swamp, and never get home again!

Also, the old woods "batches" were wise in getting the rare wild game, in season or out. They dried, canned, and salted the game and lived well, compared to the city greenies.

Though we were young and new to this Wisconsin territory, we had lived in wilder, more remote places and knew the ways to get along. We might not live like kings but we would live like free men, meeting the challenges of survival. I picked the first green shoot of the fiddlehead ferns and canned them. I gathered pigweed and canned that, too; it's as tasty and as good for you as spinach. Blueberries were rich, blue, and plentiful in the swamp. The mosquitoes drove many pickers away, but not Jim and me. We wore heavy woolen shirts, even in the damp heat. The mosquitoes couldn't drill through. On our heads we wore big hats covered with netting, and we picked until not a berry remained on a bush. We canned them all, every last little blueberry. It might be ten years until the burn-over fires spurred the swamp to produce berries again.

We had a small cranberry marsh on our place, too, and we picked a large wooden barrel full of the big pale green berries. We invited our neighbors to pick with us but they were not interested in green cranberries. We spread ours out on the upstairs floor, and along about Thanksgiving time, when they were a deep shiny red, our neighbors were happy to take them home. They learned about the cranberries, but too late. The fires killed the little marsh the next year.

Jim would spend the early mornings hunting venison, but it was scarce. Fish, likewise, were scarce. The river near our place was deep and dark, but like the rest of the country, barren. Sometimes, on clear crisp Fall mornings, I would take my little gun and wander out in the poplar grove and pick off some partridge. Perhaps a bunch of prairie chickens would settle on our clover patch and I'd take the gun from above the door determined to put one in the pot for every bullet I used. Jim might miss, but not I. That was waste. That night we'd have a prairie chicken feast, simmered in a deep pot and covered with rich cream, courtesy of our little Jersey cow.

Chapter Six

JIM GOES LOGGING

The pioneer life was not going well with most of the settlers. Some had relatives or friends who could rescue them before it was too late. They were giving up and going back to Iowa, Kansas, or Chicago, but others had no place to go; no one to turn to. So they stayed on in the grim woods, their hopelessness staring from gaunt eyes. The abandoned places were stripped of every usable thing; wire and pumps were pulled up; windows and doors taken to make repairs on some shanty.

Old JC who in the past used to sit near the window of the bank to wave a genial hand to each settler as he came into town, seemed absent most of the time now. The days of openhandedness were over. No more did he say, "Go over and tell the boys you want twenty-five cows," or "Pay me when the cream checks roll in." These days when you needed help you were lucky to see Ray Sales, JC's right hand man, and he'd say, "You got to see JC." JC wasn't in this week or the next, and so on. And then the word got around. Jake Miller went and saw JC. He said, "Pay up and that's all there is to it. No feed on time. Bring in your cows."

Well, they could all go, Jim said, but we'd stay. Jim would get in his small crop and then he'd hide behind the tie pile near the railroad. When the old stump-dodger train pulled out, Jim was no longer laughing at me from the tie pile. He was riding the rods to the wheat fields of Dakota. Then the beet fields of Minnesota 'till freeze up. Then, risking freezing to death, Jim rode the rods home again with cash for taxes, horse feed, clothes, food and seed for next year. But taxes kept getting higher and higher. Each time he thought we'd have a generous margin, the taxes would seem to gobble it up. Even the state of Wisconsin seemed intent on bleeding these people to death. Just why, I don't know. Five tax-free years would have helped them so much, and the state would have profited, too, because they were the best kind of people — not bums and ne'er-do-wells, but people who paid their bills and had saved their money, only to have it skinned out of them.

Jim now put things in order for the winter and prepared to go to work at a logging camp at $23.50 a month. He walked the fifteen miles to the camp every Sunday night through the fierce Wisconsin winter. Home on Saturday night, and back to work again on Sunday.

Jim didn't have the necessary clothes to go to work in the woods. "I'll go over to JC," he said, sure of an old friend. "I need $10.00," Jim told the seldom-seen JC, "so I won't freeze to death before I get my first pay." JC finally gave out with the ten, but said, "I want it back out of your first pay." Jim looked a little worried as he told me. "Ten dollars even looks big to JC," he said. "Can you pay the balance on the place?" I asked. Jim said, "No, I can't."

"JC is going to put the squeeze on you too, I think."

Jim's eyes said he feared I was right.

Jim left for camp that winter on December 11th. But I was not alone. Now I had little Bob with me, blonde, small and sweet, my very own little son. Snow fell early that winter, and often. It piled higher and higher at the base of the big hill that was the beginning of our road.

I took little Bob one day, thinking to visit my nearest main-road neighbor three miles away. I struggled into the snow at the base of our big hill, up to my armpits. I couldn't make it up the hill. I turned and fought my way back to the house and stayed home until April 7th. That day I put Bob into Jim's old army knapsack and plodded through the mud to the village. I had seen a woman only once in three months. She was a gabby old liar, but I was glad to see her anyway. She drove up in an outfit that looked like she had set the backhouse on runners. She had a small oil stove in there and was real comfy, plowing along through the deep snow.

We had no radio so I read everything and anything. Hundreds of those little five-cent pocket classics which Jim's brother had mailed to us. Old papers and magazines mailed from our friends in Chicago. Jim brought the mail in with him on Saturday night. These papers were never burned; they were read and reread and discussed pro and con until they fell apart. And to this day I can still enjoy reading any old newspaper. My friends chide me, saying, "Why that's last week's paper." "So what" says I, happily reading, remembering lonely days that were saved by old newspapers and their glimpse of life beyond the end of the spur road.

PERFECT
Baby Powder
PERFUMED
BORATED TALCUM
Sanitary Powder
FOR INFANTS

Chapter Seven

AN ARMY .45 AUTOMATIC FOR COMPANY

Departing female neighbors cast dire predictions on my probable fate if I insisted on staying alone in the backwoods. They recalled the spring when one young wife had been found on her cabin floor, beaten to death with a frying pan, her tiny baby beside her, frozen to death.

While I feared neither the wolves or bears, I was aware of dangers in a land of few women and many mateless males. Any man's wife, unarmed, was fair game, but I had learned how to use a gun when I tagged at my Dad's side. I was a good shot at nine years and a better one now. Jim bragged about my shooting among the bucks at the general store. He told the men that he had to give a secret whistle a half-mile from home to be sure I did not take a good shot at him. Wearing his old Army .45 when I went to the barn to do the chores, I was taking no chances on being surprised.

One night I returned from the barn, opened the door, and started into the house. A slight depression on the back of the leather davenport warned me a man's head had leaned there in the last minute or so. I stopped and stood in the door. "Hello there," I called. A big woodsman came around the other side of the door. If he had had any evil intentions, they evaporated at the sight of me with that old .45 automatic in my hand.

"You look awful dangerous for such a cute little squirt," my visitor said. "Don't you ever take that thing off?"

"Nope. Sleep with it," I said.

And I did sleep with that gun under my pillow. Somehow I could not seem to get away from this habit, and six years later when we had returned to the so-called city civilization, after ushering my brown-eyed daughter into the world, the doctor lifted the pillow and there was the gun. He and the nurse both looked so shocked that I reluctantly put my frightening old side-arm partner in the dresser drawer. After that, fear often haunted my dreams. I felt so helpless without my automatic at hand.

But to get back to the North again — I gave the old woodsman some hot coffee and a bite to eat and sent him on his way. He was nice enough, all right. I think he really respected that forty-five.

Jim's extra money from the logging camp seemed to be gone almost as he made it. My diet of bread and tea seemed to make me leaner each day. The dish of sugarless blueberries I allowed myself each day doubtless prevented scurvy, but left me a blue tongue and blue teeth. I could hardly bear to look in the small kitchen mirror.

Trying to nurse Bob when I could not maintain myself was a losing battle. Soon he became too thin. He cried all day and became blue all over, and not from berries. "He is starving," the old doctor said. So I had to buy a big jar of Horlick's Malted Milk. It cost a shocking $1.80! ! It seemed so much! How can I ever manage to pay that much, I thought; but I did.

There were other things a baby must have, too — oils, antiseptics, powder — so many things! Jim said I was becoming a real wizard with money, but I made a list of needs against the sure-to-come $8.00 government pension check. I figured the probable cost of my needs at $12.00. Too much! I took from the list the prunes, raisins, and molasses, and added it up again. Still too much, at $9.63! That would not do. Cocoa's out. That was a luxury anyway. Five pounds of sugar instead of ten. Finally I got it within my budget.

Maybe something would be a little cheaper, so I could manage a two-cent stamp. I could write to someone. I loved to write letters, and my friends loved to get them. They said I made everything sound gay and exciting. Well, to Jim and me, it still was gay and exciting, but hungry.

Chapter Eight

A LOOK OF SURPRISE

The old doctor in the village told me that now that we had Bob, we should let the Veterans' Administration know and we would receive an increase in Jim's compensation. This we did, and soon we were receiving $8.50 a month! Had I the postage to spare, I would have liked to have asked the Veteran's Administration to explain to me just how one could feed and clothe a baby on fifty cents a month.

Early that fall one of our neighbors, who had given up and was leaving, had some young pigs to dispose of. One pig, a misshapen runt, he gave to me. I made little Porky a snug house in our old root cellar. I fed him everything and anything I could scrape together. He appeared to be growing and I dreamed of pork chops and baked ham. I think there were times when Porky's stomach tormented him. I wouldn't have much to offer, and I'd say, "Porky, this is one day you will just have to put a strip of lean in your bacon." I could tell this idea did not meet with Porky's approval. He would squeal, scolding loudly. Sometimes he would break out of his pen to try and rustle up some grub on his own.

One bitter December night I gave Porky all of the night's warm, rich offering from our little Jersey cow. I took a quart of precious cow feed and mixed it in the milk to make a warm, wholesome gruel. Porky's little grunts were full of contentment that night. After he ate every bit, he crawled deep into his straw bed. Only his little nose stuck out a bit, blowing small straws out an inch or so and then catching them quickly and drawing them back against his funny snout.

With my animal family all tucked in, I went into the house where Bob and I had our supper and crawled into bed by seven o'clock. To stay up late meant burning more wood. Our stove ate wood like a hungry demon. Burning wood meant cutting more wood; cutting green wood with a bucksaw is slow, hard work. It was much better and easier to go to bed. I would wear a sweater and mittens to bed and read by the light of our kerosene lamp. It wasn't long after the wood in the stove burned out until the temperature in the room dropped to zero.

Sometime during this night, I awoke to hear sounds that seemed to tell me a very drunken man was in the little stormshed that Jim had built over our front door. The door wasn't locked, of course. No one locked a door — an old northern custom. He seemed to be stumbling about looking for the door knob. He sounded drunk, mad, and mean. I was scared to death. I thought of running out the back door to the barn and barricading myself there. I expected any second he'd find the doorknob and reel into the house. Of course, I could kill him with my pistol but I shrank from that. I looked at little Bob deeply asleep in his basket. I couldn't leave him, and I couldn't hold out long in the barn with him exposed to the cold. So I grabbed a chair and braced my door with that. I felt a bit more secure. I knew the chair would hold. He'd have to break the door in to get into the house. That would not be too difficult for any of our northwoods neighbors, but it gave me some time to plan my next move. I took my gun from under the pillow, and walked to the door. I said "Go away or I'll shoot." A cold silence. I waited, and then it seemed he'd found the door. I could tell from the sounds that his hand was near the doorknob. I knelt down. I thought I would shoot him in the legs. I didn't want to kill anyone. He was probably some good neighbor, drunk after having some home brew. I fired. All was quiet. I didn't know what to think. I was afraid to open the door. I knew now I would find a dead man there. No one left. He was there, and he was dead, and I was mortally afraid.

I crawled into bed. I couldn't sleep, but when Bob wailed his breakfast call I stirred my paralyzed muscles and made a fire. I fed Bob, then went out the back door to our barn and took care of the chores. I didn't go near Porky because I would have had to pass the front door to do that.

So I waited, afraid to look and afraid not to look into our storm entry to see if I had really killed one of our neighbors. It was Saturday and I knew that Jim would come home from camp. It was one of the longest days I ever lived through. At last I heard Jim's quick step. I rushed to the door. Throwing myself into his arms, I sobbed, "Jim, I killed a man. He's lying there in the storm shed."

"Who is it?" Jim asked, trying to calm me.

"I don't know. It was during last night. I've been afraid to look. I just waited for you all day."

Jim took me to the bed, pushed me down on it, and covered me with a quilt. "Now don't worry," he said. "Everything will be all right."

He lit the barn lantern and walked to the front door.

"Oh, Jim," I cried, as I covered my head with a blanket, "I hope it's not some little children's father."

Now Jim had the door open and held the lantern over his head to see better. With a note of command in his voice that puzzled me, he said, "Come here, honey."

"Oh, no! I can't look!"

He came over, took my hand, and led me to the door. I closed my eyes tight. Jim stopped at the open door. "Open up your eyes," he said. With pounding heart, I obeyed. I had never looked upon a dead man. I also had an uncanny fear of death. But it wasn't a dead man that lay in our storm shed. It was my little Porky pig, with a neat hole in the middle of his head and a look of surprise on his stupid little face. Tears and laughter mingled as my tense nerves sought relief. Why little Porky, with a full tummy and a snug bed, had broken out that night, I will never know. Maybe he had gotten a tummy ache from his unusually rich dinner.

I was sorry for having killed little Porky, for there went our hope for pork chops, ham, and bacon. Besides, I knew a kinship with Porky. There had been times when together we had both been hungry and cold.

The story of Porky spread far and wide, and although I was kidded about it a lot, I had nothing to fear now from any man. From then on, I was considered a dangerous woman in that part of the Chippewa valley.

Chapter Nine

HOMINY AND NOTHIN'

When the lumber camp closed in the spring, it found us with very few extra dollars to buy horse feed and seed to plant. I had hoped by subsisting through the week days on bread and tea, having a better menu only on Saturday nights when Jim came home, and then a good Sunday breakfast and dinner before Jim left for camp, to have our canned venison and greens and rutabagas carry us through the spring season. It would be the end of the clearing and planting time before we could expect any greens from our garden. This diet was strictly my own planning of ways and means. Jim never guessed I lived on bread and tea through the week, but I didn't mind. My father's house had always had an abundant table, but somehow now food was one of the lesser needs of life. One did not need much food to lie in bed and read through most of a winter.

Now, with the lumber camp closed and Jim home, we took stock of our condition and prospects. One thing was sure: If last spring was lean, this spring bid fair to be even leaner.

During the winter, Jim had purchased two bushels of very special seed corn at six dollars a bushel (a young fortune to us). But we hoped to develop a type of corn that would thrive in our short growing season. But now my days of bread and tea must end. Jim was working hard at cutting brush to clear more land. I was right at his side, piling and burning the brush as we went along. Our hardy young appetites and heavy work soon brought us face-to-face with an old problem — food for ourselves and food for the horses; for they, too, were working hard. Our mare, Bess, was heavy with foal which was due in June, and which would be as good as one hundred dollars in cash as soon as the foal was weaned.

Just when Jim should be home putting in his spring crops, it seemed that he must leave to get a job. We had hoped to get set this year so that the farm would carry itself. No farmer could ever create a paying farm when he must be far away from his fields for over half the year, earning enough cash to support his family.

That little eight-dollar-and-fifty-cent government check could not be stretched any more. The baby things which little Bob needed had to be bought. Jim and I, though packing no fat, decided to tighten our belts and tough it through.

Jim said he would give up his idea for a corn crop this year and plant oat seed which was cheaper. One bushel of our precious seed corn went to feed our team of old Maud and Bess; the other bushel of corn we made into hominy for Jim and me. "Expensive grub, anyway you take it," Jim muttered, "But it can't be helped."

Neither of us liked rabbit, but we forgot that one day as Jim suggested I take a foray and see if I could put rabbit stew on the table. I hadn't gone far when I saw a brown bunny near the chopping log at the edge of the woods. I approached rather noisily, waiting for him to run. "Never shoot 'em sitting still," Dad had always said, "Bad sportsmanship." But this rabbit acted like the original dumb bunny; just sat there. I was close enough now to see his eyes blink, except that they didn't blink. I couldn't understand it! I stepped back a pace or two and put a shot right between his ears. Nothing happened. He just sat, as dumb as ever. I went up to him, lifted him by his ears, and plunked him down. He was warm but did not even wiggle. I called to Jim. I lifted him up and plunked him again. "I've never seen anything like this, Jim. What does it mean?"

"It means that the rabbits have the sickness they are said to get every seven years," said Jim. "You'd better start supper. Hominy and nothin'."

We had started the last winter with twelve hens and a rooster. We had hoped they would produce some fryers for this coming summer. Two had died, unable to stand the terrible cold of their thin shed. The remaining ones had frozen combs and were not generally able to produce healthy chicks, so we decided to sacrifice them to our needs, and gave up chicken farming until we could give them warmer quarters. We started ending our chicken farming that night. Chicken and dumplings for supper! Then Jim remembered some winter radishes he had buried in the cellar last fall. They were crisp and hot, and just what needed to pick up our spirits.

Chapter Ten

THAT'S YOUR SILVER DOLLAR

It seemed this was to be a year when everything combined to lick us, but Jim's determination was stronger than ever. I guess I was just sticking around to see what was to happen next, even though I knew that we held a losing hand and should cut myself and Jim a new deal.

With the hominy and chicken a sure thing in our food department for a while, we fell to clearing new land to plant. After putting out our clearing fires to go in to dinner one day, we were surprised to come out of the house after eating and see our carefully hoarded stack of clover hay going up in smoke. We couldn't save the hay, but we were puzzled as to how this had happened. We found that one straw had set the next afire, and the next; a tiny thread of fire until it reached and fired the clover hay stack.

We had counted on this good hay to supplement our necessarily short grain rations for the horses. Bess, with foal, could not be fed marsh grass. Jim looked at the black ruin. "I don't know where I could buy any clover even if I had the money. When I first came here, and had money, I had all the crops insured, but that all ran out long ago."

Here was another problem! I couldn't stand the look of despair on Jim's face, so I moseyed up to the house. I got a silly notion to put on Jim's suit and hat, just to make him laugh. I'd planned to sneak up quiet-like and say, "Howdy, Mr. Welch. I'd like to sell you some good clover hay." Well, I put on Jim's blue suit with the pin-stripe, and his hat. As I took the hat from the box I noticed some papers lying on the bottom. I looked them over and found they were the insurance papers Jim had mentioned. They still had two weeks to go before expiring! I got all excited. I stuffed them in the suit coat pocket and started off mighty fast for the field where Jim was working. I could see his expression and it seemed to say, "Now, what the devil can that dude want." When I got closer he started to laugh. He sat down on a log to enjoy whatever lingo I was going to dish out. He knew I was up to some nonsense, but he wasn't prepared for "Look—you can get seventy-five dollars for that burned hay. The insurance is still good!" Jim took one look and was up on his feet and headed for the house on a lope. I trotted alongside him saying, "I remember a notice in the general store that said that clover hay was twenty dollars a ton." An hour later he was saying, "Goodbye, honey, I'm on my way. This is our lucky day after all. Hay and money to spare."

I stood reflectively in the door as Jim and the team disappeared beyond the bend in the road. I wished I were really a man. I'd put my hands in my pockets and survey the whole darned world. Nothing would ever lick me. I put my hands in the pockets of Jim's suit pants, making believe, and, ye gods! I found a silver dollar. Delirious with excitement, I was off down the road running like a deer to overtake Jim, yelling at the top of my lungs to make him hear me above the rattle of the old hay wagon. When I did get his attention, I presented such a frantic picture that he stopped the team, jumped off the wagon, and ran back in my direction. When we met I could only cling to him and gasp, "Look!" Jim's look was one of relief and disbelief. Relief that I was not reporting another sudden disaster, and disbelief that what I held before his astonished eyes, appearing to be a silver dollar, was really a silver dollar.

"Well, God Almighty! Where did you find that?"

"In your pants pocket."

"Well, that's your dollar. You go home and make out an order to Sears for those sandals you been wanting. That's just a little hunk of that silver lining I'm always telling you about. Got to get on my way so I can get home before dark tomorrow. What should I bring you from town?"

"Some jam, peanut butter and cocoa."

With our goodbyes said all over again, Jim was on his way once more and I turned toward home and the evening chores.

As I walked from the barn that evening, chores finished, I noticed a lone partridge sitting in a white birch at the edge of the clearing. I decided to take a shot at it, if it remained until I got to the house for the gun. Once I had the gun, I made an easy, clean hit. I took the bird into the house and cleaned it. I noticed peculiar white things around the spine. On close examination, I found they were worms, white as milk, thick as common store string, very alive and apparently tough, as exposure to the air and even salt water affected them not at all. I had intended to eat this bird, but now I couldn't. But I decided to keep it and take it to town and see if our young veterinarian could explain these worms. This I did, four days later, and the vet said that he had never seen anything like it. Both of us were familiar with the backwoods and we had never seen or heard of such worms. So far as I know, to this day I have never seen it again, nor do I know what caused it. The bird showed no apparent signs of ill health, but the thought of those wiry, white worms twining around my backbone changed my mind about partridge for a spell.

By four o'clock the next day Jim had returned home with two tons of good clover hay, some ground feed for our little Jersey cow and the horses. Jim also bought the peanut butter, cocoa, and jam I had asked for, besides sugar, flour, coffee, prunes, raisins, tea, and even a book of stamps so I could write to my friends. The final prize was three months' supply of Bob's special baby food.

Chapter Eleven

CHARLIE DAVIS' BIRTHDAY PARTY

By the end of the week we had turned our new clearing of about three acres, gave it a quick spring-toothing and tossed on the rutabaga seed. A couple of light weed cultivations with the spring-tooth after the "beggies" were up would be all that was needed to produce the one sure-fire settler's crop. The rutabagas could withstand the frosts which we could expect almost any month of the year. All the stock could eat them and thrive. For the stock we chopped them up with a small hand-ax and sprinkled a little salt on them. Frozen or not, it did not seem to make much difference to our stock. Mostly the "beggies" were eaten frozen.

For family use, we might serve them a different way for each day of the week, and then serve the raw slices or strips with salt in the evening, making believe they were apples. Some people just hated the things in any way, shape, or form, but we liked them. Thank God for that.

It was nearly the middle of April now, but not too late for a good heavy snow. It started snowing early one evening and by ten o'clock there was two feet of heavy, wet snow on the level.

Two miles to the south of us lived Sam Towns, a one-time Virginia coal miner, and his eighty-year-old mother. She was Grandma Towns to everyone — plump and jolly and beloved by all. She was one settler who could sit tight, having a widow's pension of seventy-five dollars a month. She was considered to be very rich, and she was also generous and wise. There were many little children who lived through those first winters because of Grandma Towns' beans and spuds, not to mention some of Sam Towns' out-of-season game that helped fill many a pot in the cabins along the spur road.

Grandma Towns lived with her Sam in a little, black tarpaper shanty that was snug and warm. Grandma and Sam were both so friendly to all the new settlers. It seemed like Grandma always had so many good things to eat at her shanty. I swear she must have just spent all her time cooking, hoping that some hungry settler would come by so she could stuff raspberry pie or venison stew into him. Meanwhile, Sam would entertain the guest with naughty stories or he might play his banjo and sing old hillbilly songs. Sam could play the banjo fine. He knew the words to heaps of old songs, but he knew only one tune. His voice sounded like he was going to cry until he would break out in a great roar at how awful he was and say, "Hell's bells, I wish I could sing!" So everyone, as often as they could, found their way to Grandma Towns' place. We sat on the beds; we sat on the floor; we came there blue with cold and stomachs aching with emptiness. We left there gay, our stomachs full — aching now because we'd laughed so hard!

Well, the night of this heavy April snow, Sam hiked over to our place and told us that if the snow held over through the next day, he would like to take his mother and some of us neighbors over to celebrate his Uncle Charlie's birthday. We'd all go in Sam's sleigh, about ten miles, right across country.

Old Charlie Davis, Grandma Towns' brother, was seventy-five, tall, lean and good-looking. How he remained a batchelor all his life, I will never know. He also played the banjo and guitar. He sang those old mountain songs with such charm! As he was a favorite in the neighborhood, we were thrilled at the thought of surprising him on his birthday.

The next morning Sam was along about eight o'clock, the hay piled deep in the sleigh; hot stones here and there, and plenty of blankets and deer skins. I put a pot of beans in a corner of the sleigh with some hot stones around to keep them cooking. Little Bob and I snuggled close to use some of the heat, too, blankets tucked tight up to our chin and around our feet, and we were off across the back country. We needed no roads. The forest trail was clear and beautifully white with fresh deep snow.

Sam's team was young. Wild, too, because he hardly ever used them. They reared and snorted and nearly pulled Sam's arms off. He "Whoa'd" and begged until he would get mad and say, "Bessie, you damned fool!" all of which made no difference to Bessie at all. When we pulled up at Uncle Charlie's door in the late morning, surprisingly without mishap, Bessie was still acting like a steam engine about to blow into pieces. However, her harness partner, who was her mother, was nuzzling her, seemingly saying, "Simmer down, little one." Charlie stuck his handsome head out the door when he heard the team jingle into the yard. He looked more like thirty-five than seventy-five. He was surprised. I never saw anyone so surprised or so happy.

We trooped in, glad to get the kinks out of our legs from sitting so long. We put our food out on the table, and what a spread it was! Charlie made coffee and then disappeared down into a hole in the floor and came up with about six kinds of jam. He lived near the Moses marsh. You could go into the marsh and, if you could get out alive, you'd have some nice berries for jam. But what a hell-hole it was — mosquitoes big enough to carry off a horse. It looked like Charlie had made a few forages into the marsh. Charlie, too, was an old soldier with a pension. His life was lonely, but free from the usual money worries.

We ate; we sang; Charlie played us a tune on his saw. He made a wooden doll dance a jig on the saw blade. He sang songs in his wonderful way, while he played his guitar, and he looked at me with his deep black eyes. I wished, somehow, that this party would last forever. Everyone else could disappear for all I cared, except my little Bob. I wanted to stay in this cabin alongside the river bank forever, with this charming man singing his gay songs to little Bob and me.

The spell was soon broken. The air was warming fast and bare spots were beginning to show on the ground. We had to say goodbye to all our friends and hurry home before the snow was gone. As we rode home I said to Jim, "That is the sort of place I have always wanted. A log cabin on a river, far away from everyone else, lonely and quiet and peaceful."

It had been a wonderful day for everyone. We were not downhearted by the sudden snow. It had done no one harm, and winter was as over as it ever gets in the north, which is to say it's only around the next corner. This had been just a harmless little joke of nature and it provided us with a fine holiday.

LITTLE GEM

Chapter Twelve

THE STOVEPIPE WAS CHERRY RED

The day after the party dawned bright and fair with just a spicy nip in the clear air. It was hard to believe that yesterday was a winter day with two feet of snow. Jim said the snow was invaluable to the crops. It did not melt and run off the ground; the warm earth just inhaled the moisture into every pore. "To the clover, this is the nectar of the gods," he said.

We were in a rush to take up our farm work again the next morning so Jim cut a big basket of pitch-filled pine chips from a large dead pine stump that edged our clearing. Pine chips made a quick fire in the cookstove. I gave Bob his morning bath in front of the open oven door. Jim had never put so many chips in the stove at one time before, but when he said he'd quickly make a good hot fire, he sure did! In no time at all the water was warm and little Bob was in his tub. Jim had gone to the barn to get the horses ready for field work. I was hurrying with Bob's bath to be ready to join Jim and make a tour of the back forty. We wanted to map our next job of clearing fields. The stovepipe was soon cherry red, but that was not unusual. However, the smell that came into the air was. Behind the stove the heat had buckled the wallboard out toward the stovepipe, and the pipe, being now red enough to melt, soon set fire to the wallboard. That tarpaper shack would burn fast! I never was the type to scream in an emergency. Screaming always seemed like such a waste of time! The thought of yelling my head off made me feel like an ass. I didn't scream now. Jim should be within hearing, and could come to help me, but I went into action on my own. I jerked Bob out of the tub, wrapped him in a towel and laid him on the bed in the next room. I ran back to the kitchen and threw the bath water on the burning wallboard. Most of the water splashed to the floor in a big puddle. Quickly I soaked a turkish towel in the puddle and slapped it against the burning spot. By this time Jim had the horses harnessed and came in the door to see how we were getting along. The minute he opened the door and saw what was happening, he reached behind the stove, grabbed the edge of the buckled wallboard and tore the burning area out, and threw it out the door. By now the fire in the stove was dying down. All the fire danger seemed past. Jim looked around.

"Where's Bob?"

"On the bed in the other room," I answered. But we went into the next room and there was no little Bob in sight. The door had been closed, no one had been around, and Bob certainly couldn't walk. He could wiggle a little but he had never gotten off the bed alone. We looked under the bed. We looked all over the room. He just wasn't there!

"You were so excited maybe you put him down in the kitchen someplace," Jim said.

"But I didn't. I know what I did. I put him right there, naked and wet."

Then it seemed I could hear him. It sounded like he was inside the wall near the bed. It was a silly idea, but that's how it sounded. The bed was set right up tight to the wall. I crept to the head of the bed carefully to listen for this muffled sound. There I saw him! He had wiggled over to the wall and had fallen down between the bed and the wall. He couldn't fall to the floor nor, of course, could he get back up on the bed. He was just small enough to show neither below nor above the mattress and bedding. Well, the poor little fellow was cold and blue and almost smothered, but we soon had him dressed and comfortable. With everything under control in the house, we got to work in our fields.

Bob was such a sweet little fellow. Long before he could walk he joined us in everything we did. Jim fashioned a small box on runners and put this sled behind the harrow. I lined the box with a big blanket. Little Bob trailed along behind the field harrow, where he bobbed happily until he nodded to sleep. When he was asleep we took him out of his sled and put him in an old army hammock that Jim had set up at the edge of the woods, safely screened with a big army bed netting. There he would sleep for hours, to awake bright-eyed and fresh, ready to be tucked into his dad's old army knapsack, for a ride on Jim's back as we went on a hike of our rocky domain. If we decided to stop at some spot to do some work, we simply hung the old knapsack on some convenient tree or pine stump root with never a fuss from little Bob. He was a real pioneer baby and a joy to our lives. He was always happy and gay.

I dreamed, as all mothers do, of Bob's growing up to be a fine man some day. He would need many things as he grew — clothes and food; and he'd want a bicycle, skates, baseball mitts, and so many things that are a part of every boy's life. Most of all he would need an education. He must play college football. Whatever Bob needed, or wanted to be, we must help him. This was to be our true destiny. I felt this was why I was born. God had given me this wonderful little bit of manhood because He knew I would fight to help him to be a credit to the world.

Even supplying Bob's simple baby needs was a problem. It was soon brought home to me that we could not give him security while living here in this barren Wisconsin wasteland.

Chapter Thirteen

A LIFE FOR A LIFE

Although Bob had been born with a rupture in his groin, we were not aware of it until he was about nine months old. Seeing this sudden change in his small body sent us hurrying to the village doctor who lived eight long miles away. The doctor said the rupture had been present since birth but only now had made its appearance. "Doc" Stone was a medical reprobate. His father had spent his life in kindly service to the poor of this dreary country. It had been Old Doc's dream that his son should carry on in this helpful practice. Young Doc had finished medical school and returned home so that his dad might retire. I guess young Doc hated the guts of every bushwhacker in the area; at least he handled them like he did. He had a beautiful wife and I think he resented not being able to give her a better life. I think Old Doc's heart must have ached when he saw what he had wished on his people.

One winter evening, Old Doc sat talking in the office, admiring the new spaniel pups that had just arrived. These pedigreed dogs were young Mrs. Doc's source of clothing money. This particular batch would bring her $300.00 for her Christmas shopping. The door opened to admit Lennie Karns. He had been running, and couldn't talk for a moment. He looked at Old Doc as if to say: "If I could only ask you."

Old Doc had brought Lennie into the world and had done the same for nearly all the other old settlers. Of course, many babies had come into the world unattended. They told how Grandma Sams' only reaction to delivering a child was that she was late with supper — a late supper ten times in her life! She greeted her husband on those nights with, "Little late tonight. Your new son arrived this afternoon." No disruption in the household routine, really — just a slight delay.

Young Doc and Lennie Karns had gone to grade school together. Lennie looked at him now with a pleading to remember those old days.

"Ray," he said, "You got to come right away. It's Sara Jane. I think she's dying."

Young Doc poured himself a whiskey.

"Lennie, I'm not going to go out tonight for no damn kid — yours or anybody else's. I'll see her in the morning."

Lennie's face went white. He knew young Doc. No use cursing; only bring on a cursing, and such a scene laid heavy on old Doc. Lennie turned and stumbled out. How could he go home and face his family and say Doc wouldn't come?

Old Doc said, "I'll be running along. Here come Chris and Emily." Chris and Emily lived with old Doc and took care of him and his house. When they came into town to buy groceries, Old Doc usually came along and sat a while visiting with young Doc.

Emily said no good came of those visits. Old Doc was always depressed afterwards. He wasn't proud of his son. He'd just get in the car for the return ride, and never say a word. He'd just eat his supper and go to bed. But this night Old Doc seemed to spring into the car. He turned to Chris and said, "I want to go out to Lennie Karns' place."

"Sure," Chris said. "Right after supper."

Old doc replied, "No, I want to go before supper. I want to go now." Emily looked at Old Doc's usual calm, impassive face. It was flushed and excited.

"You'd ought to stop by home and take some more of your medicine," she said. "Your cold isn't all well yet."

"I can skip a dose, I guess," Old Doc snuffled. "Little Sara Jane is sick and I want to see if I can do anything for her. Ray wouldn't go until morning. Said no kid with a bellyache was going to keep him from taking his wife to the Saturday night dance. Well, I'm too old to dance and I've got no wife, but I won't sleep good 'til I see that little girl of Lennie's."

When Old Doc opened Len's door and barged in, followed by Chris and Emily, they found Lennie trying to explain to his wife, Nancy, that young Doc would be there early in the morning. A big smile broke through her tears when she saw Old Doc come in. After he had warmed his hands and examined Sara Jane, he came out to the kitchen. He seemed to be deep in thought.

"Let's all have some coffee," he said.

Nancy put on the coffee for all of them. They drank their coffee and watched Old Doc. No use prodding him. When he decided what he wanted to say, he would say it, and not a second sooner. At last he looked into Lennie's face. "Len, it's diphtheria and I don't think she'll live through the night unless her throat is lanced."

"Then you've got to lance it, Doc; there's no other way."

Chris and Emily said, almost in unison, "Be back with your bag in half an hour."

Old Doc looked at his hands. "I'm old and rusty, Len, and shaky."

"I'm not worried, Doctor," said Nancy. "I feel like God sent you to us tonight. I know you're going to save Sara Jane."

"Well, you know, Nancy, I feel sort of sent, myself. Just seemed I was hell-bent to come here. I know if I'd thought it over I'd never have had the courage to come."

Len tried to get him to eat some supper while they waited for Chris to come back, but Doc said there was no time to waste eating. They would use that time to get ready. Nancy had a new kettle which was soon ready with boiling water to be used as a sterilizer. Doc's hands were scrubbed and re-scrubbed. Doc took off his old flannel shirt and he put on one of Lennie's Sunday white shirts.

Chris was back in a little less than the half hour with the medical kit. Now, no one spoke save Doc, giving orders, but everyone prayed. At last Doc said, "That does it." Great beads of cold sweat gathered on his forehead, then a whiteness around his mouth. Doc made as if to turn and walk away, and his knees buckled. Len and Chris caught him and laid him on the bed.

Old Doc slept all through the night, quiet as a baby. Chris and Emily sat up with the child while Len and Nancy got some rest. Little Sara Jane's temperature was near normal by morning. Her struggling breath was now restful and quiet and she slept most of the time, a smile on her sweet little face.

Old Doc seemed to be resting after his ordeal, a smile on his face too, but with a temperature of one hundred and five. This, young

Doc discovered when he showed up at ten in the morning. He was surprised and a little ashamed; although on young Doc the latter emotion was hard to see. Old Doc wanted to return home, but as it was thirty below zero they persuaded him to wait until the outside temperature went up and his temperature went down.

Three days later the outside temperature was fifteen above, but Old Doc was dead. Len and Nancy felt so bad. Len said, "He just gave his life for Sara Jane." Emily said the day before he died, Old Doc had asked her how Sara Jane was, and she told him "Just fine." She said Old Doc seemed to know he was going over the hill. He said, "Emily, I call it a fair exchange; a mighty fair exchange."

Chapter Fourteen

JIM RIDES THE RAILS

There were those who had known young Doc for a long time, who said he seemed to change after his father's death. Everyone thought he'd leave the country. He'd always acted like he hated it and everyone in it, but now that he was free to go, he stayed on just for pure cussedness, some said. Others said that young Doc couldn't make the grade where folks would have a choice of doctors. However, it looked as if he was staying on and becoming, at times, almost human. He certainly had mellow moments those later days.

It was in one of these moods that I found him when we took little Bob in for an examination. The last thing I wanted to do was cry in front of that cranky, hard-boiled Doc. Although I gritted my teeth and tried to squeeze my eyes tight, the tears splashed down on Bob's little blond head for Doc to see. He just started cluckin' like an old mother hen. I couldn't believe my ears. I looked up to see if it was really Doc.

"Now, here, honey, don't you cry. That rupture can be fixed. He'll be all right. I'll call Dr. Vinson in Ladysmith right away." Mrs. Doc came in at this point. I think she wanted to see if she could believe her ears, too.

While Doc got busy making the call to Ladysmith, Mrs. Doc gave Bob a little toy. "To take home," she said, and she told Jim and me not to worry — everything would be fine. By now young Doc was back. Dr. Vinson had told Doc that Bob was too young for the operation; that he should wear a truss until he was about a year and a half old, and then the doctors would do the operation. We paid young Doc about $2.50 for a truss and went home with young Bob, a new truss, and a new problem.

The operation for Bob's rupture would cost us about $150.00, Doc had said. Then add to that the hospital expenses and other costs. It all seemed impossible, but a way must be found. We had nearly a year's time to plan, but time wasn't money in this country. It wasn't anything except just time. Jim said not to worry. He would go where the money was, as soon as the crops were in, and he'd bring back the money we needed, and more besides.

First on the plan was to follow the harvests and work in a pea cannery, but the pea crop wasn't so good. Some soured and were wasted due to humid weather and transporation snarls. Then, on to the Dakotas and the harvest in the wheat fields. Jim did fair, but the I.W.W.'s were stirring up things out there and put obstructions in the path of an honest worker. Then on to the Minnesota beet fields. Jim had been over all this ground before, but this year the weather conditions were terrible — half-freezing rain, wading in cold muck to your knees; hands wet and cold all day long, and then before you could make a cent, a blizzard ended the whole thing.

Jim climbed between two freight cars and hung onto the ladder. His hands were frozen numb; there were times when his mind blacked out; he thought he had dropped to the rails and had been ground to pieces under the wheels. But somehow he came to his senses to find his hands still on the ladder. He told me later that he didn't expect to ever see Bob and me again. He knew he was more dead than alive. He remembered another fellow had taken his place on the ladder just behind Jim. Jim suddenly realized he was gone. Jim thought he felt him move down and under — not a word or a sound. His face would have been too frozen to speak or cry out, as was Jim's own. Jim wondered whatever Bob and I would do. What would become of us? We didn't have anyone except each other. Jim's family life had ended when he stood one day, a lad of nine, in patched, clean overalls, his only pair, beside his mother's grave. Jim was sent to the Indian school at Carlisle, Pennsylvania. His father had been killed on a hunting trip when Jim was only five.

He thought about my childhood — The day I was two months old, I lost my mother. Dad, crazy with grief, had taken me from our Oklahoma home to an aunt who lived in Illinois. Dad followed the ever-fading frontiers, to be heard from maybe here, or there, but never for sure from anyplace. Jim didn't want to leave Bob to the half-orphan life he had lived.

Finally, Jim felt the train slowing down. He thought that if he could unload soon, he might find a hobo jungle nearby. If he was still on the train when they stopped in the yards he'd land in jail; so he leaned his whole body to the left, jumped, and pitched head-first down a ten-foot embankment. His face plowed easily through the snow into the sharp cinders, but the smart of his burned face seemed to him warm and good.

Jim was never sure of the time or events that passed between the time he hit the embankment, and the time he walked in our back door. He said he remembered the last time feeling warm and full. He had been at a trackside hobo camp. There was a big pot of stew, and strong coffee. He felt it was a dream. He closed his eyes and slept again for hours. When he awoke it was early morning. The fire in the camp was burning low but the coffee was still hot. He drank a couple cups of coffee and walked about a little to limber up his legs and hands. Yes, he felt like he could start out again. A covered bundle nearby moved. A rough face stuck out and inquiring eyes looked at Jim. "Take it easy, boy," the bundle said. "I'm O.K.," Jim said to himself. To the bundle he said, "I'm on my way. I want to thank all of you. Here's five dollars for the pot." The bundle took the money, said, "Good luck," and my Jim disappeared in the shadows of the dawn. He knew he was in no condition to bum his way now, so he bought a ticket and coached the rest of the way. Jim was still a mightly sick man when he got home. I put him to bed and fed him up on fatted goose — a goose we had received a few days before from an out-going settler who owed us a small debt. I put goosegrease and turpentine on Jim's chest. But his ears turned black and flopped down like an old boiled cabbage leaf. Jim's feet were too sore to walk on.

Chapter Fifteen

AN OLD BLUE BUCK

The Bolton Brothers came over to talk about Jim going on a pre-season deer hunt, but he wasn't interested. He was far from recovered from his recent experiences and the chances of his bringing back a buck were too slim.

The year before, the Boltons and my Jim had put the jump on the deer season. The net result was an old blue buck that Mother Bolton had ground through the finest knife of her food chopper to make mincemeat. Even with Mother Bolton's treatment, followed by a long, gentle simmering, the finished product was like spiced bits of rubber bands.

Even though Jim declined to go hunting with the boys that year, their visit was most welcome. We played "High, Low, Jack, and the Game" until long after midnight and laughed again over last year's hunt.

Jim and the boys had put a new tent, a tiny sheet iron stove, blankets, ax, guns and a few pots, a lantern, grub, and who-knows-what into Bud Bolton's old Ford, and set out to get their deer. They wanted to beat the invasion by city slickers. The native boys did not favor roaming the woods at that time. In spite of flaming red clothes, when a city fellow goes forth to hunt, well preheated against the Northern cold with his favorite brand of spirits, anything that moves is a deer to him. And enough times for discomfort, he makes a direct hit. So it was the custom for the native sons to get their venison a little early, and then either act as a guide to Mr. Tenderfoot or hibernate till he returned to the city.

The boys stopped, well-coffeed, and told the local game warden about the probable site of their camp. He could then make a wide circle of their hunt and avoid seeing them. They proceeded on into likely deer country to arrange their camp before dark. The tent went up, with a stove in the middle to give them warmth. Supper over, the boys went out to cut marsh grass, then piled it up to the eaves inside the tent and trod it down well. On top of the grass, choice balsam boughs were laid. Each man arranged the boughs to please his fancy; then another layer of marsh grass, topped with blankets. A wide bare circle was left around the little stove in case a spark escaped to start a fire. After gathering wood for morning, the men decided to turn in for the night.

"Cook breakfast before daylight," Bud told Russ. "Be ready to start hunting at the crack of dawn."

"What time would that be?" asked Russ.

"Oh, I reckon if we get up at five," Len replied. "What time you got now?"

Everyone looked at everyone else and then they discovered there was no timepiece among the four of them. They had a good laugh on that one but it didn't worry them any. They all said they could tell when it was time to get up without a timepiece, even if it was dark.

In a little while the rumblings from the little tent would have scared any deer, bear, or wildcat right out of the country. Bud woke up first. He cooked coffee, bacon, and eggs.

"Come on, you deerslayers. Sun'll be over the marsh here pretty quick."

They turned out of their warm beds and ate. They waited for daylight to break. While they waited, the men fired the little stove. It soon burned out. They fired it up again, and it burned out again.

"I'll be dadburned if I believe it's more'n two o'clock right now," Len said, "I'm turnin' in again."

So they all turned in. Some time later Russ stirred; remade the fire, more coffee, and kicked the other three out of their beds.

"Come on, you guys. Look alive there. All the deer will be leavin' their marsh beds and headin' for the tall timber."

Jim poked his head out the tent flap. "Russ, it's darker than Hades out there."

"Yes, I know, but it's always darkest just before the dawn. Didn't you ever hear that?"

Well, they'd all heard that, all right. It was a good argument. So they stayed up and drank the hot coffee. Bud said they'd better open a can of beans, but long after the beans were gone, darkness still remained.

Each of the hunters had taken a turn at getting the rest of them up, making some more coffee, and opening some food, before daylight.

The dawn, advancing on them so silently, found them all sound asleep. A restless morning breeze stirred the tent flap enough for the sun to hit Russ square in the eye. That did it. They were up and away in a few minutes, to find the steaming, empty beds of the deer families. The hunters played hide-and-seek with the deer among the timber most of the day. No shots were fired all during the hunt. A white flag here and there, but just a flash and the deer was gone.

Toward dusk, they returned to the marsh, thinking the deer would soon emerge from the timber to bed down in the marsh, but the deer were alert and stayed in the timber. All, that is, except that old blue buck. He thought he was safe because anyone would know he was too old and tough to eat; but at the sight of him Bud's gun went up.

"Hold it," Jim said. "Lord, God! He's twenty years old! What the devil you want with him?" But Bud had drawn a bead on a vital spot. The old 30-30 brought the buck down.

The men stayed two more days, but no deer were located. It was annoying not to know the time and place for the hunt.

No, Jim didn't go hunting again. In the Spring, Bob had to have that operation. We had to have the money for it; we had to talk and make our plans. This was no time to waste on possible deer meat.

Chapter Sixteen

I'M HIRED! I'M CAMP COOK!

Jim was glad he turned down the hunting trip. A few days later, Fred Steadman came along and said he had a job of road work to do, not far from us. Steadman wanted Jim and our team to work for him and he would pay five dollars a day. What a break it was for us! I hoped that the job would last for days on end, but it lasted just eleven days. How big and timely that fifty-five dollar check looked! It proved to me there was money in this world, and I was going to find out where.

I asked Jim if he couldn't work for Fred at some other place. Jim said Fred was finished building roads for this season. "Going to have a logging camp up on the Jump River," Jim answered me.

"Maybe you could work for him in the camp," I said.

"Yes," Jim said. "He wants me, and Maud and Bess, too. He'll pay forty dollars a month, and for the team another thirty, with grub included for all of us."

I thought Jim's plan was wonderful, but I also knew every camp must have a cook, and I determined that I would be the cook at the Jump River Camp. Just what wages a cook was paid, I did not know, but I knew they got more money than anyone else in camp, and that the woods boss would rather lose anyone else in the camp than to lose his cook. I'd ask fifty dollars a month in wages. Fifty dollars plus

Jim's forty, plus thirty for the team! Everyone would eat good. No living expense. Why, by Spring we'd be all set to have Bob operated on for that darned rupture. I was so excited by the thought that I could hardly speak.

"He'll need a cook, Jim. I'll be the cook. Fifty dollars a month. Add that up, Jim. That's IT, Jim! That's the answer to our problems! See, Jim — it's really simple!"

Jim looked at me as if he thought I had suddenly taken leave of my senses. He thought I was playing a game. Then he realized I was serious.

"Look, honey, that's a big job. And you've got Bob to take care of, too."

But in the end I convinced him, or thought I did.

"Big cookin' just runs in my blood, Jim. My Uncle Bert was head chef on the Santa Fe. He used to tell me how they do things."

"I never knew you had an uncle," Jim said.

"Well, if I HAD had an uncle on the Santa Fe, he would have said that if you can make one loaf of bread, why can't you make a hundred loaves? If I can bake a pie, why can't I bake more than one pie? If one man eats a pound of meat, a hundred will just eat a hundred pounds of meat. Simple arithmetic, Jim. Instead of making a cup of tea, I'll make a big pail of tea or a barrel of tea."

I had never worked at a paying job, and the thought of asking for a job turned my knees to knocking. Although I seemed to have convinced Jim that I had the know-how, I wasn't sure that I could convince old Fred, so I made Jim promise he would ask Fred about the cooking job the very next day.

"Don't forget," I told Jim the next morning. "Tell him just like I told you, and remember the part about Uncle Bert."

All day I figured and planned in my mind. In my mind I cooked huge stacks of flapjacks. Very simple. I just took pancake mix from a big barrel. Nothing to it. I made dozens of big fat pies; loaf on loaf of warm brown bread; gingerbread cakes; cookies; doughnuts, brown and sugary. By night time I could hardly wait for Jim to come home. I knew Fred would have been glad to get me. Jim would have told him what a wonderful cook I was and that I was young, strong, and quick, and also that I'd never go to town, get drunk, and forget to come back, like the men cooks did sometimes.

At the first sound of the team and wagon on the road, I took Bob in my arms and started out on a run to meet Jim. When I thought he could hear me, I started to shout. "Did he say yes?"

Jim stopped to take Bob and me aboard the wagon.

"Well, hurry up! What did Fred say?"

"Fred say?"

"Yes, about my being a cook."

"Gee, honey, I just plumb forgot to ask Fred."

I knew Jim was lying. He hadn't forgotten at all. He didn't believe I could do it. He had hoped that it was just a crazy idea and by now I would have forgotten it, or had gotten cold feet.

I was quiet the rest of the way home. I was more determined than ever. I hugged little Bob tight, and told myself I COULD do it, and that I WOULD do it, for my little Bob.

I didn't try to sell my idea to Jim any more, but after supper as we sat talking, I said, "Jim, promise you will ask Fred tomorrow. I mean, truly promise."

Jim's brown eyes looked into mine. He wanted to talk me out of the whole thing, but the determination in my eyes was more than he could buck. He just said, "I promise."

When he came home the next night, it developed that he had not asked Fred if *I* could cook for him. It seems he asked him if he *had* a cook.

"No," said Fred. "Have to look up a man somewheres."

So Jim let the matter drop there. It seemed apparent to him that it was a man's job. He had never heard of a woman cook in the woods. Now I lost all my fear of asking Fred for the job. I was good and mad. No, I had never heard of a woman cook in the woods, either, but so help us, a woman cook there was going to be in the Steadman camp. I didn't say much to Jim that night. I guess I "mostly pouted," as Jim would say.

When I took Bob's truss off that night, I held the noxious thing up and said, "In the spring, Bobbie boy, when you get your operation, I am going to throw this thing as high as the moon."

"I hope you can," Jim said.

"Well, I will."

"Listen, honey," Jim cautioned, "don't count too much on this Fred idea. You see, he wants a man. The thought of a woman cook has never entered his mind."

"'Forget it," I said, as I crawled into bed. My plans for the next day were all made.

Old Fred had the reputation of being a sourpuss — wary where women were concerned. But in his middle forties a saucy seventeen-year-old girl had come to visit his camp. She chewed her gum in the old pirate's face and answered his growls with sass, until one day Fred looked at her long enough to decide she was a darned cute little dish. Now she was home, busy taking care of their six kids. After that, Fred had a soft spot in his heart for brash young girls; so the next day when he looked up to see me standing before him, with Bob tucked in the old army bag on my back, his eyes widened with interest.

"Mr. Fred, meet your new camp cook."

As old Fred started to glance around I said, "No, it's me, Mr. Fred. I'm a good cook. I can do everything in the kitchen — bake bread, pies — cakes — anything."

Old Fred took my arm and led me to a sunny spot along the roadside. He looked so flabbergasted he was glad to sit down. I started right in telling him all about my Uncle Bert, and all I had learned from him.

"Well," Fred said, "it's a pretty big job. I'll have about thirty-five men, and they eat an awful lot. Pancakes for breakfast, and meat and potatoes, prunes, cookies and doughnuts. Then pie every day for dinner, and baked beans, too. Bake all the bread. Yep, it's a big job. You're still in your teens, ain't you?"

"Oh, no, Mr. Fred, I'm in my twenties now," I lied.

"Well, now, what wages you going to ask?"

"Fifty dollars a month."

"Well, that's all right, but I'm danged if I can see how you can do it. You're no bigger than a pint of cider. What if I'd decide to have a bigger camp — say fifty men?"

"Oh, Mr. Fred, I wouldn't care. Fifty, sixty-five, or a hundred, I can do it. I'll make a deal. If I do the job up proper, you pay me a fifty-dollar bonus. If I flop, you don't pay me a cent. Is that fair enough?"

"Oh, that's more'n fair. But they say it's bad luck to take a woman into the woods. That's why you never see women cooks. And you're too young and too pretty. Even if you were old and fat it would still be dangerous. I'm afraid to take the chance. I'd be uneasy all the time."

"Mr. Fred, I'll make every man-jack afraid to look at me, even out of the corner of his eye. I'll wear my old .45 day and night. Jim can tell you I'm a crack shot. I've stayed alone back here in the woods weeks on end. No man dared to come on the place. Please, Mr. Fred."

Old Fred closed his eyes with a sigh of resignation. Then he opened them, looked steadily at me. He arose, reached a hand to pull me to my feet, and then, still holding me by the hand, he said, "Mrs. Jimmie, I can't argue you out of it. I guess you don't scare worth a darn. You're hired!"

I was so happy I couldn't talk. I put my arms around old Fred's neck and gave him a big kiss, then dashed off down the road to tell Jim "I'm hired! I'm the cook!" — and completely forgetting little Bob. He had wandered off away from Fred on little-boy business of his own. When I went back and scooped him up to put him in his army sack, Fred was still standing there, his cap pushed off to the side of his head, scratching his ear and saying, "Well, I'll be dadburned!"

Chapter Seventeen

KEEP YOUR EYES ON YOUR PLATE!

Wild with the joy of getting the cook job, I didn't have a worry in the world. Jim was to leave for camp with the horses on October fifteenth. About ten men would be there to build the necessary buildings, and they would cook for themselves. By the first of November they should have the cook-shack and bunkhouse finished, and then they would be ready for Bob and me — their new camp cook.

Until then, I spent my time putting our affairs in order so that we would be ready to leave. I gave the little Jersey to a neighbor to keep for her milk; loaned our heating stove to another, and listened to the neighbors on every hand telling me that I was crazy to think I could cook for that bunch of men. Why, it was unheard of. My husband was crazy for letting me, and Fred was the craziest for hiring me. All this fell on deaf ears. I guess it was just a case of not knowing it couldn't be done, and going ahead and doing it.

I had prayed that God would show me a way to earn the money we needed for Bob's operation, and here it was. Maud and Bess's wages would take care of all our needs, and when spring came, we'd draw out our combined wages and take care of all our money problems. I'd get that bonus, too, for our special needs.

At eight o'clock, November first, Fred and Louis, his blacksmith and general right-hand man, drove up to our farm. Bob and I were all set to go, but I still had two members of our family that I wanted to take along, and yet I hesitated to ask Old Fred. One was our dog Shep. She had reluctantly stayed behind with Bob and me when Jim had left with the team. She loved to ride on the wagon tongue between the horses, and they had never before left without her. She *could* hunt her food in the woods and sleep in the haystack, and no doubt be home to greet us in the Spring, but she would miss us and the horses. She had stayed with me these two weeks and I didn't want to leave her alone now. As Fred and Louis loaded our things, Shep tagged about like she was trying to figure it all out, and wondering where she fit in.

"Your dog?" Fred asked.

"Yes," I answered.

"Well, don't look so sad." Fred opened the car door and said, "Get in here, Shep. You're goin' loggin', too."

"Are we all set now?" asked Louis.

"Cept for our pussycat. I promised little Lonnie I'd take care of her."

Louis hauled a gunnysack out of the car and went into the house and scooped up Pussy, and we were on our way.

Thirty-five miles north they stopped the car and told me that the path through the woods at our right led to camp. Since the car had to be left at Madden's, I started slowly toward the timber while the men put the car away and passed the time of day with the Maddens. I soon found the path and, with Bob in my arms and old Shep following, we were off into the timber. About a half mile down the path I came upon the camp. It was sure bare; not even a fire in the cook-stove and it must be getting near noon. I put Bob in one of the bunks that lined the room and soon had a fire going.

Suddenly Fred and Louis came panting up. They had run all the way. Missing me back at the car they thought that I was lost in the woods.

I said, "You can't lose me in a woods. But where's all your pots and pans?" They both roarded with laughter. "Why, this isn't our camp! It's a mile further on. Get your coat on. Jim will be getting' worried."

So on we went — Fred and Louis taking turns carrying Bob, Shep and I bringing up the rear. At last ahead of us was a clearing in the white pine forest, and there was the camp. I loved it at once.

Set in horseshoe arrangement was the cook house, joined by the bunkhouse. Then, without a door and commanding a view of all the goings or comings, was the "backhouse." There one could sit and view and be viewed. All activities at the barn, blacksmith shop, office, and general store were plainly visible to whoever occupied the throne at the moment. If nothing of interest was taking place in the areas before him, the time could be spent going over the various catalogs at hand.

Before entering the campsight proper, you had to cross over a creek about fifteen feet wide on a little old tree which looked about six inches wide. I knew the bets were all on my falling in, and I never was good at that sort of thing, but darned old Shep walked out to the middle of the log and did a fancy ballet, turned around and came back to me, just to show me how easy it was, I guess. I decided if she could manipulate four legs out there, turn them around and come back, I ought to get over on my two legs, with arms as balancers besides. So I just walked smartly across, and looked up to see Jim's hand ready to grab mine. "Hi, pardner," we both said at once.

Jim had been elected temporary cook of sorts, so we sat down to his meal of potatoes boiled in their jackets, boiled beef, prunes and biscuits, coffee and jam. The crew, which now numbered ten, was occupied with building and putting the camp in order for operation. Then they would become full-fledged lumberjacks, to be joined by fifty more.

Jim's feelings weren't hurt a bit when the men said they were glad of the prospect of getting some "real cookin'." He was glad to get out into the fresh air to work and come in to my cooking at night. Old Fred introduced little Bob and me to the crew.

"Men," he said, "we got a real lady here. And this little fella. We don't want him to learn any bad habits from us so I don't want any goddam swearing in this camp."

The men could hardly keep straight faces. Fred hadn't realized he'd been swearing. It was that way with all of them — it was part of their picturesque way of expressing themselves. Unless a man said these words in anger, it wasn't considered cussing.

As soon as the dinner dishes were out of the way, with little Bob at my side we explored our surroundings. I had never been in a logging camp before, and was fascinated. The cook house was something of a shock to me, for I was bashful. When the men poured in, I wanted a little corner to hide myself in, for when so many pairs of flashing bold eyes looked at me, my face turned a bright red, but as the cook house was planned, I was in full view at all times. Fred couldn't stop those bold looks the men gave me on entering, but he stood watching until the last man was seated. Then he said, "Keep your eyes on your plates. You're in here to eat." Occasionally Fred looked down the line of backs. A few times he saw a face turned my way with an intent burning look. He got up, grabbed the bold one, and jammed him down on the bench until it seemed his spine would be shoved through his skull, and it would be his last meal with us. Fred would follow him up and pay him off.

Chapter Eighteen

THE COOKHOUSE — MY CASTLE

Old Fred's camp soon became known as a good camp. The cook was young and pretty and nice; the food was fine and the cook's little boy was a fine little fellow, but old Fred brooked no monkey-business. No sheep's eyes at the little cook, and no booze. If Fred caught a man entering camp with a bottle, he turned him back in blizzard weather, forty below zero, even if he was drunk and likely to die before he reached town. Fred's office was situated where every man must pass by on entering camp, and Fred's eagle eyes were usually on him a good while before he opened the door to pass the time of day with him. Sometimes he handed him a check. "You're through," he'd say. "Wait right here. I'll send for your belongings." So, for the most part, the bums and the wise guys gave us the go-by. We became a camp of good family men. Now and then some undesirable got in but he was soon sent on his way.

The layout of the camp was extremely primitive compared with a modern city — but so workable! No waste space; no frivolous un-essentials. The cook house was a long rectangle, running north and south. At the south end was the cooking and dishwashing unit and the cook's bed. Two windows in the entire building — one opened to the south and one to the east. Unless a window could catch the sun, it had no reason for being, in lumberjack logic. In front of the south window was the sink where the dishes and huge kettles were washed. You could pour the water into the sink to be carried off by the nearby creek. This was a great convenience. Standing at the sink washing dishes, you had the big cook stove at your back, not six steps away, handy with its huge thirty-pail reservoir of hot water. Above the stove was a rack where the hot pots and pans were placed to dry after scalding — no wiping needed.

The table dishes used by the men were enamel plates and bowls. The bowls served as cups, soupbowls, or oatmeal dishes, sauce dishes, or anything you wanted to use them for. These, upon washing, were stacked in racks and set on the stove to dry, and then reset on the table for the next meal.

After they were washed, the steel knives, forks, and spoons were put into a large flour sack. With the cook holding one end of the closed sack, the helper holding the other, they shook it back and forth several times, emptied the contents out on the table, and they were dry.

At the right of the sink was a huge wooden box that looked like it was meant to be Paul Bunyan's casket, but the inside of the grub box was divided into compartments for bread, cookies, and doughnuts. The contents were always available to any wayfarer who happened by. You needn't ask him to partake; he just went over and helped himself; took a cup of coffee from the big pot that constantly simmered on the back of the stove. If you chanced into another camp, you did the same. It was an old logging camp custom.

The end of the grub box came flush up against the end of the cook's bed, which was built into the southwest corner. The end of the table, which could seat fifty men, came tight against the side of the cook's bed. So every night, Jim, Bob, and I climbed onto the grub box, walked over the foot of our bed up to the head, and down under our covers. The big table ended at the other end of the cook-house, with only just room for the men to pass around to the far side of the table. The whole north end of the cookhouse wall was lined with shelves bending under their load of gallon tins of food and twenty-five pound boxes of prunes, rice, macaroni, and other staples.

Coming back to the east side of the cook house, beginning with the other side of the stove and in front of the east window was a huge breadboard, handy to the stove oven and with a long shelf above the window on which to put my pies to cool. At the left of the mixing board was a table on which to put the finished cookies and bread to cool. Beyond this extended a smaller table set for twenty-four men.

In the middle, between the two tables, was a great, long, flat-topped heater, the kind you could load with cord-wood length logs — regular log cuts that kept a good fire throughout the night. More often than not, Jim awoke in the night, made the trip to put a couple of logs on the fire, and all was very toasty when I got up at 5:00 A.M.

The bunkhouse and cookhouse were perhaps twenty feet apart, joined by their roofs, making a passageway between with east and west sides open. Here was a chopping-block to cut meat and hooks to hang the carcasses high away from bears and wolves.

The bunkhouse had a small window at the north end; a door at the south. A large sheet-iron stove stood in the middle with a rack built over it where the men hung their sox to dry. The resultant potent atmosphere seemed in no way to harm the honest sleepers who occupied the double row of bunks that lined both walls.

Between the bunks at the north end, in front of the small window, a table was built, and with the building's one and only chair a jack might read or write if he could manage to see in the dim light. At the right as one came in the door, was a wash basin, water, and a mirror, if one were vain enough to need one. Each man provided his own toilet articles — towel, comb, and razor if he wished. No inner-spring mattress here! A new arrival went to the barn, got enough hay to fill his bunk, got blankets from the office, and that was it. Pillows were not furnished, but you could roll your coat under your head if you felt the need of one.

At the left of the bunkhouse door hung the teamsters' lanterns, all clean and bright and with the driver's number in sight on a wooden tag. In Fred's camp there were ten teams, ten drivers, and the lanterns hung in order, beginning with number one.

Number four could not get up and go to the barn first! No; Number One was head man; no one moved until he did, and then only in order as they came, so if Number One was lazy and a laggard, late to the barns, it followed that all must be late. But the teamsters were looked upon as super-beings and they really seemed to be. They had to get up an hour earlier to care for their horses. They couldn't go from the supper table directly to the bunk-house to loll on their bunks; they must go to the barn to feed, curry, check and care for their charges. To their credit, I can say they took pride in their work, and the lead teamster never needed prodding in the morning. No lazy man would choose to be a teamster, even for the slight increase in salary. Being a teamster involved much responsibility, aside from the extra hour morning and night, and it meant only good men need apply. But, even among lumberjacks, there were men who loved horses. Among the ten teams, none could compare with our Maude and Bess for beauty, strength, and temperament enough to be a challenge to a skilled teamster. Many men expressed the desire to drive and work with our team.

Old Fred had closed the door on applications for cook's helper by announcing one day that he was fed up with every-other man asking him to be my helper. When we had a crew of twenty-five, Jim would be "cookee" to Mrs. Jimmie, which meant we must have a driver for our team. Anyone interested in that job could talk to Jim, Fred said. So it began. It seemed everyone wanted that job, but they did not know that team like we did.

"Maud must choose her own driver," Jim told me. Every man who wanted to take on the job had a tryout. He drove around the circle of the campsite with Jim and me watching quietly from the enclosure between the cookhouse and bunkhouse.

"See, Maud is surly," Jim would say. "She would watch for a chance to kill him. And Bess is nervous; she'd soon bite a chunk out of his arm. Can't take a chance with those devils."

So it went, day after day. No good. It looked like we might have to board them out, idle all winter. Bess was heavy with foal, and while exercise was important, she needed careful handling.

One day, as Fred and Jim stood before the hitched team, an old neighbor of ours, Jed Barton, came around the corner from the woods with his two sons, Edgar and Wesley. They had never been in the woods to work before, they said, but thought they could take their paycheck without blushing for shame. Edgar knew our team, having visited our place several times. He went up to old Maud and stroked her neck; he scratched under her blond mane. Maud liked it; she nuzzled his arm. Bess seemed jealous. She pushed Maud aside to get Edgar's attention.

"So you like me, you pretty gals," Ed said.

Fred looked at Jim. "They haven't shown any affection for anyone around here so far. Maybe there's your driver."

"Just what I was thinking," Jim replied. "How about driving the ladies, Ed?"

"Lord, I couldn't; they'd pull my arms off," Ed begged off. "Besides, I'm green in the woods."

"Don't let that worry you," Fred said. "If you can manage the team, the rest is easy."

So Edgar drove the team around the circle, cut into the logging road, a mile out and back, with Jim and Fred sitting in the back so quiet Maud never knew they were there. She obeyed Ed's every order as though it was a pleasure, and Bess was calm and easy. When they came to a stop in front of the bunkhouse, I could see that Jim and Fred were pleased, and Ed was certainly more confident.

"There's your man, Jim. You're Number One teamster," Fred said, turning to Ed.

Chapter Nineteen

BOILED TEA AND SOURDOUGH CAKES

Not only was Louis, the blacksmith, Fred's right-hand man, he was everyone's friend in time of need. He knew Fred like a book, which in itself was an accomplishment. It was well to learn early if you had a problem to take up with the boss, it was wise to put your case before Louis first. He would advise you how and when to approach the old sourpuss, if he advised approaching him at all.

While I had told Fred I had learned much about cooking from "my uncle Bert," it soon became apparent that there were some things I had not learned. I don't suppose "Uncle Bert" had ever had to cook without eggs, not even one egg all winter. Fred had not told me this; Louis said he had not considered it important as he was sure at the end of the first week I would be ready to go home. But I did not intend to go home until the last man had gone down the trail at breakup. This, Fred was beginning to believe, also.

He said, "We usually make a couple barrels of sauerkraut. Don't suppose you can do that."

"Why of course," I said. "Just get the cabbage, salt, kraut-cutter, and barrels." One Sunday afternoon, aided by many willing hands, we cut, salted, stomped, and tasted two barrels of kraut. Fred was mighty pleased.

"But I bet you can't make mincemeat," he said, thinking he had me stuck there.

"Well, I just can," I sassed. And the next Sunday the air was spicy with the odor of cooking mincemeat. The men loved to spend their Sundays at such jobs, and it was a great help to me, and I sure had risen in Fred's estimation. But there was one thing I could not seem to do — make pancakes without eggs or fresh milk. I tried various combinations. They ate them but they said they weren't right. It seemed no one knew the secret of sourdough pancakes.

Before I confessed to Fred that I just did not know, I went to Louis. He tried to help. He got a cookbook of hundreds of eggless recipes from his wife — a book put out during World War One. It was helpful in many ways, but on pancakes the result was not what the men wanted. Finally, I told Fred. I asked him about buying pancake mix. He laughed. "Fluffy-duff for sissies," he called that. "He'd ask his wife," he said. He and Louis made a trip home over the weekend. They came back with a recipe for eggless pancakes.

"You make a starter, like for bread," Fred told me. It sounded pretty vague. They admitted that they did not understand it themselves, so they could hardly be expected to make it clear to me. Fred was nice about it. Very patient. But he said, "It's very important. The men will complain. It's like beans and prunes — they're a MUST in camp life."

I took a two-gallon crock and put in potato water, sugar, a little salt, and plenty of dry yeast. It was in the nature of a ceremony, advised and approved by Fred and Louis. Yes, that was the idea, they thought. You set it away in a warmish place for a while "to work." Just how we would use it from then on wasn't quite clear, but we set it in a cupboard beneath the sink and beside the stove.

Some days later I remarked to Jim that there seemed to be a bad odor someplace. He sniffed around and agreed. He said he would give me a hand at cleaning that afternoon when no one was around. It would look bad if the men should consider us dirty. So far, no one had seemed to notice the smell but us. It didn't take long to find the trouble. Jim opened the cupboard door.

"It's in here," he said. In a moment he lifted out the "starter."

"It's this stuff. Wow! What a stink! I'll get it out and dump it in the creek before anyone sees me."

So I had the pancake problem back at zero again. While Jim was at the creek washing the jar, I stood glumly wondering how I could ever find out how lumberjack pancakes were made, when Louis walked into the kitchen.

"Say, Mrs. Jimmie, that pancake starter ought to be about powerful enough to use by now."

"Powerful? You don't know the half of it. Jim is just throwing it out, it was stinking so. It was awful!"

"Ye gods! You threw it out? That was just getting ripe enough to use."

Now we had to start all over again, and when we produced the next big stink, we were on the first step to success, and I made hundreds of luscious big brown pancakes that met my lumbercamp family's critical approval.

My camp family was growing every day. We soon had twenty-five men at work. Jim was my helper and he did many things that an ordinary helper would not have had to do, such as kneading the huge batch of bread dough, or getting up ahead of me to get the fire good and hot, and having the coffee made and the potatoes frying by the time I had to turn out.

Old Dad Barton was my bull cook, and I know there never was a better one. The reservoir was always full of good hot water. He never let it get lower than a couple of pails at a time; therefore our water supply was always adequate and hot when we needed it. Also, Dad kept the drinking water always fresh and cold. Our water was supplied by cutting a hole in the ice above the point where the sink drained into the creek. Dad said he didn't mind carrying a half dozen pails at a time, but to face the job of hauling thirty or forty pails at a time would make his back ache just thinking about it. Besides, he liked to have a good excuse for dropping into the kitchen often, as he enjoyed a cup of coffee with a slab of pie now and then.

Dad Barton kept the woodbox full with all of the best wood — just the right mixture of dry tamarack for a quick morning fire; good seasoned maple for the baking fire, and a sprinkling of green stuff to give a slow fire when I wanted that.

Dad kept the bunkhouse ship-shape, too; the lanterns all cleaned bright and shiny; plenty of fresh water to drink and two pails of hot water on the flat iron stove, that a man might need for soaking his sore feet.

Dad Barton was seventy-five years old, a big and bony man. He got to his work early in the morning. So well organized did he have his many duties that he seemed to have none at all, but I doubt that many men half his age could have done as well. No sir, Dad Barton did not need to blush with shame when he drew his pay!

One day Dad said to me, "Mrs. Jimmie, the boys mostly like your cookin' fine; the pancakes are gettin' mighty good, but one thing they complain of is that you don't make good tea."

"Can't make tea? Well, gosh, what's wrong with my tea? Jim and I drink tea at home all the time. I thought I made good tea."

"Nope, it ain't right. Been a lot of men talkin' about it."

An hour or so later Fred came in. "Mrs. Jimmie, I know you learned a lot from your Uncle Bert, but why in hell didn't he teach you to make tea?"

Now I knew it was serious. Dad had tried to tip me off before Fred would have to come to me.

"Mr. Fred, I don't know what it is. I thought I made perfect tea."

"I don't know what's wrong, but I know something is. Maybe you need to make it stronger."

"I thought it was awfully strong now, but I will make it still stronger." But although I made it stronger, it did not please them. I went to Louis in tears.

"Well, I don't know," he said, "but they sure are griping about it."

"Gee, I never heard of more than one way to make tea. I think the guys are nuts."

"No, they're not. It isn't right."

I went to the cookhouse feeling pretty low. Imagine saying I couldn't make tea!

That evening, as I was about finished with the supper preparations, Louis dropped in. He stood around "chewing the fat" as they call it. I threw the tea in the pots — about six times the amount I really thought enough. I poured boiling water over it and set the pots on the back of the stove. Louis looked from the tea pots to me and back again in amazement, and exclaimed, "Don't you boil it?"

"Of course not. You don't boil tea."

"Oh, yes, you do. That's what's the matter with it."

So I boiled the tea. I could not drink it. I told the men it would make them very sick; it was now a poisonous brew. But they only said, "Ah, that's what we call *tea*." And they were all happy again. How could Uncle Bert ever have known about *boiling* tea! !

Chapter Twenty

MY WHITE BIRCH OUTHOUSE

The early fall weather had been remarkably fair. Fine for building and putting the camp in order. Everything was about ready to put into actual operation as soon as we had a good snowfall. Then the water wagons would go into operation at night to turn the logging roads to icy ribbons.

Old Fred was more than ready for a sharp change in the temperature. Fifty men lined the tables three times a day, eating their heads off. The crew, too, was primed and anxious to hear old Fred sing out "Lay the timber low, boys!" There were puttery jobs to keep a man busy, if he had a mind to do them, but they longed for man's work.

The boys decided that the cook should have her own private privy and that idea seemed wonderful to me. As matters now stood, I had to make a trip before daylight and before any men should be stirring; then again while all were seated at the table, for I could not bring myself to answer nature's call in full view of anybody and everybody.

Fred agreed they must do this for me, but each day it seemed to be put off for some other pressing matter, until one day the men just up and refused to budge. I heard Dad Barton say, "No sir, Fred, it's going to be done today."

I paid no attention, not knowing it concerned me. About eleven o'clock, Old Dad came to the door and said he had a surprise for me. He led me behind the cook house and there was such a cute little outhouse of white birch poles! A sign on the door said "MRS. JIMMIE. PRIVATE." Well, gosh, I was kind of embarrassed, but sure pleased — it was so cute. I said "Thank you, Dad" and gave him a big kiss. At that, the rest of the crew came around the end of the bunkhouse with shouts of "Hey, he didn't do that job alone." Fred emerged with a big grin. "Well," he said, "maybe we can get down to work now."

Chapter Twenty-One

SUNDAY IT'S LEFTOVERS

Winter soon took over and every part of the business of making logs ready for the sawmill went into full swing. The biggest table in the cookhouse had forty-eight men seated there three times a day. The smaller table, planned for twenty-four men, was not always full. Sixty to sixty-five seemed to be average. No man liked to sit at the far end of the table, so it was arranged that no man would sit there. Starvation point, it was called. The men used to get a man there if they disliked him, and starve him out. Keep food out of his reach, and play deaf to his calls to pass this or that, until the unpopular one gave up and left with a bad case of nerves, as well as a big streak of lean.

We arranged our tables with a food set-up in the middle of an area of table space, bound by three men on each side. That is, in the middle of six men square. There was a complete meal menu; Salt, pepper, coffee, oleo, meat, vegetables, bread, pie, so spaced all over the table, and it was impossible for any man to be cheated of his full share. Sometimes the kinds of dessert, such as pie or cake, might vary, and these were traded around to meet likes and dislikes, but by this time no man would go hungry if he didn't get his favorite pie.

I soon knew almost to the pound just how much of this and that my "family" would eat in a day. For instance, one day we served pork — one-half of a hog; the next day, beef — one quarter of a beef; 25 pounds of lard, 25 pounds of navy beans, a bushel of potatoes, 50 pounds of flour, 15 to 20 pies, a bushel of cookies and doughnuts, 6 to 8 gallons of vegetables; 25 pounds of prunes, 10 pounds of coffee, 2 pounds of tea, 10 pounds of oleo, 25 pounds of white sugar, 25 pounds of brown sugar. I made their pancake syrup. These were the staples, day in and day out. There were raisins, rice, macaroni, dried apples, etc., in varying amounts, used here and there, too. Also a keg of pork sausage used to stretch out the meat, or 100 pounds of weiners at night meal with sauerkraut.

The boys said they never had had such good food or such variety. I was always trying to think up something to surprise and please them. Not having any eggs, they often longed for something that "looked eggy" as they put it. One night they sat down to a yellow, creamy rice pudding and golden Johnny cake. They certainly were puzzled, and old Fred most of all. I never told anyone my secret except Jim, but here it is, and it's so simple I wonder no one thought of it before. We were not allowed to color the oleo, so I saved the little yellow capsules that came with each package and used them to make certain foods look like I'd put eggs in them. They pestered me for scrambled eggs for breakfast, but I said we couldn't afford that.

Old Fred said, "I know you haven't got any eggs. I guess by now you could teach your Uncle Bert a trick or two." But I just blushed and smiled. I knew Fred didn't believe my "Uncle Bert" story, but I wasn't going to admit that I'd been such a liar.

We worked out a program now that made for ease of kitchen operation. Of course, each day except Sunday we had to bake bread. Then I had to make biscuits or Johnnie cake to get through. They didn't eat quite so much on Sunday, because they just laid around the bunkhouse relaxing. Everyone had pie for dinner except Sunday when I had only whatever leftover pie there was on hand, and I made fresh golden cake, sponge cake they thought it was, and some spicy gingerbread. One day we made cookies, molasses ginger cookies, round as a saucer and thick and rich and SO good. A real man's cookie.

The day after cookie day was doughnut day, and on Sunday, neither of these. We had found this plan would take care of our family and maintain a reserve in our big grub box for stray visitors.

Pancakes were a must each morning, six days a week, as well as potatoes, meat, prunes, cookies and doughnuts. On Sunday morning I gave them oatmeal in place of pancakes. Baked beans were a noon meal must six days a week, but these I skipped on Sunday. This dropping of all tradition on Sunday caused some grumbling at first. Fred attempted to take the men's part, but I held firm. "Sunday should be different," I maintained, until they came to look forward to Sunday as a day of some special surprise. There might be hot biscuits with honey or jam at breakfast, or big fluffy "yellow" dumplings for dinner, and meat pot pie for supper. Just like home, they'd say.

Everyone and anyone was welcome in the kitchen on Sunday, and no one wanted to sit idle. They peeled potatoes for that day and the next. They ground pork; we made pork sausage for the next Sunday. We tried ideas and recipes from this country and others. "My mother makes these," or "My wife makes these Sundays at home," they would say — and I'd make it too!

Chapter Twenty-Two

KITCHEN HELPERS

The two best kitchen helpers were brothers, George and Ed. Edward was twenty-three, married, and had a son about little Bob's age. George was twenty and hoped to be married at the end of camp. He had ordered the engagement ring from Sear's catalog and when it came he showed it to Jim and me, when no one else was around. He asked permission to hide it on the eaves above the grub box. Whenever the three of us were alone in the cook shack, George would hop up on the grub box, take the ring down and look at it. "Gee," he'd say, "I wonder if she'll like it. Well, come Christmas and I'll find out." He could hardly wait.

We'd talk of girls and life, married life; how to make a living, and such, and while we talked we worked. George was wonderful. Whatever I might be doing, if I turned away, George took over and things went on just the same. Ed was equally handy but more bashful. He thought George made a pest of himself.

One Sunday after dinner George and Ed told Jim and me to put on our coats and take little Bob and get outside. It was a lovely day. They said they would do the dishes. When we came back, the dishes were finished and the table set. Everything was perfect and two big cakes were just being taken from the oven. The boys looked so pleased, like it was the most fun they had had in a long time. They said they'd like to team up with me and get a bigger camp to cook for next winter. They wanted to help me this winter and learn. They figured they could make more money and they liked it better than their present jobs outside. Jim, on his part, would not want to spend another winter in the kitchen. He preferred outside work.

I had only Bob's operation in mind from this winter's labor. I hoped the next winter to be able just to stay home and enjoy my young son. But I wondered how it happened that George and Ed were so handy in the kitchen. They told me their mother was an invalid and, as they were the only children, it fell their lot to do most of the cooking and housework. They tried to make good things for the little mother upstairs in bed, and they'd make pie or cake and take it up to receive her praises. If it wasn't just right, she would point out the possible errors and they would try again — many times, if need be, until they gained perfection.

With the idea in mind of being a camp cook, George now spent every possible minute in the kitchen, much to Ed's disgust. Once George insisted on having a try at pie-making. The result was good; in fact, no one would have known that I didn't make them, except that it was Sunday and many had seen George at the job, and George bragged to all and sundry that he made the pies today. At mealtime George waited restlessly for someone to taste the pie and come forth with a compliment. I expected it, too, and was prepared to see the buttons pop right off George's shirt. Ed took the first bite. He looked at George. I knew he thought George was getting the big-head. Ed ate the bite of pie slowly. George waited expectantly. Ed's face gave the impression that he was too thrilled for words. Ed was enjoying George's suspense.

"Come on Ed, say it's good. It is, isn't it?" begged George.

But Ed laid his fork on his plate and pushed his plate away and turned to me. All eyes were on Ed now. Even old Fred loved a good joke, and Sunday was the one day he permitted gaiety at the tables.

"Mrs. Jimmie," said Ed, sneaking a sly look at George's face, "Could I have some of yesterday's pie?"

A roar of laughter filled the room and George, red-faced, left for the woods, but after the men left he sneaked in again to join Jim, Bob, and me at our dinner. When we assured him it was good pie, he felt better. All the men ate it and liked it — even Ed. He just had to tease George.

Ed was serious, though, about George making a nuisance of himself in the kitchen. He had talked to him about it and would call him out on various excuses, but George would be right back again. This Sunday, as was their usual custom, the men took their blankets outside to air. Ed had called to George several times to come and "shake our blankets out." George would say "Yes" but didn't go at all. Ed looked in to the cookhouse now and saw George back again. Ed thought we'd like to have our meals alone as a little family group.

"Come on, George, help shake your toots out of these blankets." Ed's voice was loud and stern. For the second time in an hour George left the cook house with blushing face, to be seen no more that day until supper time, then entering very subdued and leaving right after eating. Ed must have given him an awful talking-to.

Chapter Twenty-Three

A CAMP CHRISTMAS

Most of the meat to feed the camp was bought from the surrounding settlers, butchered at their farms and brought into camp ready to cut up and cook. Two cents a pound for pork; three cents for beef. Prices for meat seemed mighty low. It WAS low. These farmers were not much better off than we were and they could produce only low-grade meat. It was the first time I had ever eaten tough pork. We got some that winter. It fried up about like wet leather.

One day I looked up from cutting cookie dough to stare into the big brown eyes of a little Jersey cow. While still wondering how she got there, Shorty Newberry, who had ducked down out of sight, popped into view with a big smile on his face. I rushed out. "Did Fred buy a cow for us? Gee, we can have milk and cream for our coffee!" (Oh, wouldn't that be next to heaven?)

"Well hold everything, Mrs. Jimmie," said Shorty. "Old Fred bought her for meat, but he decided she's too skinny. Said we'd fatten her up a little first."

Well, she was such a nice little Jersey cow. We named her Rosalee, and now little Bob had fresh milk to drink and we all had cream in our coffee instead of canned milk, which was sure fine. We found she liked the cold pancakes from breakfast, peelings, and just about anything from the kitchen. She grew fat. She was young and so sleek; it was really too bad to have to butcher her. Finally one of the men wanted to buy her for $75.00 when camp broke up, and Fred agreed. Fred had paid $30.00 for the cow and her feed cost was very little.

When we came into camp there were two subjects Jim and I often talked about. One was the subject of bathing. Most of the men agreed they indulged in a bath only twice a year — Fourth of July and Christmas. They often had to skip Christmas, they said, because it was too cold. It was agreed that to bathe once a year was enough for any man. We were a little concerned about this program, fearing the men might get lousy, and then little Bob, for he was with them most of the time. He met each teamster when he watered the horses at night, and then he'd take a firm grip on the lines some distance behind where the driver held them, and drive them to the barn. When the first team came in, if Bob wasn't out there a shout would split the air: "Come on here, Bob. How we goin' to get the horses in the barn without you?" If Bob should be toasting his toes in the kitchen or bunkhouse, he'd hustle his little short legs out and take a good scolding for "not being on the job." Then he'd go into the bunkhouse and visit till suppertime; then, often, back again till bedtime.

Once when I mentioned to old Shorty about hoping the men were keeping clean and I didn't want Bob to get lice, Shorty said, "Don't worry none. Fred always has a clean camp. Got to boil our underwear out every week, and besides, we got citronella."

"What's citronella?" I asked.

"You mean you come to camp without any citronella?" asked Shorty. Taking a bag from his pocket and showing me some fine brown powder, he pulled back the neck of my dress and shook some of the powder down my back.

"I'll get you some next time I go for grub. Never have lice if you use citronella. I'll give Bob a dose, too."

The other much-discussed subject was whether or not you were going home for Christmas. When camp had first gotten underway, it seemed no one was going home for Christmas. "Won't see me out of the woods till Spring" was the common comment. But I noticed, as the day came closer, an air of indecision and longing seemed to follow the question.

"You going home for Christmas?"

"Well, I don't know."

"My Mom says she's looking forward to seeing me then."

"Well, you know how it is. The wife and kids."

"How about you, Fred?"

Fred had given the impression he would consider any man a sissy that had to run home for Christmas. But now Fred looked a little sheepish.

"Hell, I don't want to," he said, "but the wife and kids are raising the devil about it. Guess I'll have to go."

It seemed that, after Fred's remark, more faces took on a "me, too" expression. I told Jim they were just trying to be tough. They all wanted to go home. At the beginning of Christmas week it seemed twenty-five men were staying at camp, but at Christmas Eve we found just ten of us there, including Jim and me and little Bob, so we decided to make a real Christmas. The boys cut the top off a balsam tree for a wonderful Christmas tree for Bob. The men were always whittling things out of the soft white pine, and there now appeared a wooden chain, twenty-five feet long, to wrap around the tree, and then wooden fans, little wooden cages with wooden balls inside, wooden cupids and wooden tools. I fashioned gingerbread men, with raisin eyes, cranberry mouths, and coconut hair. Also cranberries strung with popcorn. We sure were busy, Sears' catalog had been our shop, there were toys for little Bob, mittens, caps, scarves, and plenty of candy.

Little Shorty set off Christmas eve to see if he could find a turkey for our Christmas dinner. He came back with a fat goose and two chickens. The next morning we were up early to prepare our feast. I told the men to help me and we'd make ice cream. We took a big wash-tub down to the creek, filled a gunnysack with ice from the creek, then crushed the ice with the flat side of a double-bitted axe. This was put into the tub and mixed with plenty of salt. Meanwhile, I had made the ice cream mixture, and put it in three one-gallon pails. Now we embedded them in the ice of the tub and then the men took turns at swishing the pails by their handles, this way and that. I opened the pails a couple of times to stir with a spoon and, of course, taste. At last the job was done: crushed ice two inches over the pails and set the whole works aside till dessert time.

Yum! yum! It was a good Christmas dinner. Ten of us at the table. We ate so much and we still had a lot of ice cream left, so we decided to treat the oncoming crew and we were so glad we could, because everyone came bearing gifts for little Bob and me. They had all thought, "That poor little boy won't have any Christmas way up there in the woods."

"My little girl sent this."

"My little boy wanted Bob to have this. He was afraid Santa Claus wouldn't find such a little boy way up in the deep woods."

Bob was so excited. It was the most wonderful Christmas I ever knew. Over fifty rough old woodsmen coming in bearing gifts. Some of them hiked over thirty miles down an icy river to bring us these things. An orange "inside my shirt so it wouldn't freeze." An apple "next to my skin, Mrs. Jimmie. Only way to get it here." A small baking-powder can with six fresh eggs broken into it "for Bob's breakfast." A pretty hankie for me — "My wife sent you that." Candy. A pocketful of nuts.

Old Fred said he was afraid we hadn't had any Christmas — until he sampled our ice cream. "Well, I'll be gol danged," he said.

We sang songs — the Christmas songs, and then gay songs, sad songs, many I had never heard before or since: *The Chatsworth Wreck* and the sad lament of the guy who said "if the wind had only blown the other way." (It seems that if the wind had only blown the other way he might have been a single man today, but sad to relate, the wind had hoisted a long skirt to reveal a shapely leg and the gent had been a "gone gosling" right then and there, to his everlasting regret.) I sang *Peggy O'Neil* which they always liked. Then our attention was called to the east window of the kitchen. There it appeared that four suits of long winter underwear were putting on a ludicrous ballet. They had flapped in the icy breezes just long enough to have taken on some very funny poses. Now they jumped up and down, swung and swayed, until we about split with laughter at their crazy antics. Then Shorty, George, and Ed and Jim appeared beside each dancing figure to take a deep bow. Fred said that should be the "Grand finale" so we all said "Good night and a Merry Christmas."

Chapter Twenty-Four

GABBY SPRINGS A SURPRISE

The day after Christmas dawned clear and sunny, a hint of spring in the air. A hint much too strong for Fred's peace of mind. After breakfast that morning he spoke to the men: "Boys, I don't like this weather. It looks like an early breakup. I'm asking you to dig in. Beginning tomorrow we have breakfast fifteen minutes earlier and supper fifteen minutes later, and as the days get longer further changes will be made. We've got to use all the daylight while we got ice for the logging roads. Anyone belonging to a union can step out now."

No one protested. The men, thereafter, left the breakfast table and disappeared in the darkness fifty feet from the doorway, instead of starting in daylight. They were to be in their working areas when daylight came upon them, and they stayed there until darkness settled down at night, except for riding in for the noon meal with the logging teams. Some of the men were getting so far away they complained they didn't hear Jim's mess call he blew on his old army bugle; therefore we got used to straggling late-comers. Fred hinted they had better take a lunch, but the men said they'd quit if they had to eat frozen sandwiches for dinner.

However, whenever any man was missing from the table, everyone felt a tenseness, fearing an accident. So far we had been very lucky. Although Fred's announcement of a longer day had seemed to set well in the morning, the men seemed restless at noon meal, like something was on their minds. They looked often down the road toward town. I mentioned it to Jim: "Maybe they been thinking it over and are going to hit the trail," I said.

"Well, I don't suppose the swampers care for the idea much, but a sawyer with a good partner can make a nice bonus," he said.

I had visited the skidway on Christmas day. It looked to me like all the logs in Wisconsin must be piled there, ready to start on their way to the sawmill, but Fred said he wanted to make that pile look like a kindling-pile beside those he hoped to stack there before breakup.

Now Fred noticed the stalling about, too; men in little groups talking, looking down the road; looking at me, then at Fred; indecision on their faces. Instead of ordering them to "hit the ball" as would be his custom, Fred seemed to decide to wait a little. He also took to watching the road — for what, he didn't know. Soon two men hove into view; walked like they were very tired and realized dinner might be over and were giving a last spurt in hopes of getting in on some hot food.

Fred glanced at his men. They saw the newcomers too, but it seemed that's not what they were looking for. Soon three more men were in sight. The first two were almost up to Fred by now and nine men were coming up the road. Fred was mighty puzzled. He looked back at his crew; only amazement there.

"Hold it," Fred said, moving forward. "What brings you here?"

"The word's out in town you need a hundred men here."

"Well I don't, and you can high-tail it right back."

"My God, man, give us something to eat. We've walked eighteen miles. There's fifty men behind us."

By now there were the nine newcomers around Fred, tired and angry.

"What'd you pass out the word for if you don't want men?" one asked.

"I didn't pass out the word."

"Well, the guy said you sent him to pass it around you wanted a hundred men the day after Christmas."

"Wha'd he look like?" old Shorty spoke up.

"Young; about twenty-five; wore green corduroy pants, bright green, and brown pullover sweater like a college kid. Said he was married, though, but his wife was rich."

"That lousy so-and-so" broke in Louis. "That's Gabby. He ain't back since Christmas yet."

"Well," Fred said, "that is just a smart-aleck trick. I'm sorry, but you can't come into camp. I want no strangers now. I got smallpox in camp just this way one winter and ten men died. I don't want to take chances. You men hustle back and turn your comrades around before they have to walk as far as you did."

Jim and I had been taking in the scene from the kitchen window. I felt so sorry for those poor men, but Fred said it was the only way to do.

Fred turned to Louis and told him to go out and take his car and head for town — get the word around that the Steadman camp had no need for men, now or later, and if he saw young Gabby he was to knock hell out of him. After some whispering together among the crew, old Shorty stepped forward.

"Fred, there's something we got to tell you and Louis before Louis starts out."

And then it came out. Gabby Harns, he of the green pants, had suddenly appeared in camp late one Sunday afternoon about the middle of November. He proceeded to tell one and all that he didn't have to work; he had a rich wife.

"Yessir! My wife just inherited a thousand dollars," he would say impressively.

The young men put him down, pronto, as a big windbag, and were not slow about telling him just what they thought about a gent that thought he didn't have to work because his wife had money. Admittedly, anyone in that country who possessed a thousand dollars was out of this world.

Gabby's treatment was anything but gentle. No one anticipated his staying long anyway, and if the general atmosphere should shorten his stay, so much the better. But it soon appeared that Gabby apparently could take it and that he might well make quite a stay. He was good for nothing but a swamper, and as not too many of the men were swampers, even a green swamper was welcome. Fred said, "When you get sick of his talk, just tell him to shut up, but give the kid a break. We need swampers."

Well, they did. He seemed to learn and sobered down. Didn't talk so much or so big, especially after the news got back from his home town to the effect that his wife was a nice girl who worked as secretary at the paper mill. Seemed unlikely she was rich even to the extent of a thousand dollars. Gabby got awful quiet after these things were called to his attention.

"Yep, maybe you don't have to work," little Shorty said, "but it appears your wife does."

Gabby seemed to manage to get to town most every Saturday night. He, along with a half dozen other young bucks, would start off right after supper to hike the eighteen miles to town to dance. They usually caught a ride part of the way in and back. This was not Gabby's hometown, and from the talk, one gathered that Gabby let the village girls assume he was unmarried. We did not think knowledge of this would have caused Gabby's wife any concern. Jim and I strongly suspected that his complete absence from the home fire was a great relief to her.

However, early in December Gabby had approached the men in the bunkhouse one day about all going together and buying Mrs. Jimmie a Christmas gift; not just some foolishness but a real fine gift to show their appreciation of her good cooking and her friendliness to all of them. Well, the men thought it was a good idea and asked Gabby, being a guy with a rich wife, what he proposed to donate. Thereupon, Gabby said he would put $10.00 out of his money on a forty-dollar wristwatch the next time he went to town. This he did, bringing back the receipt to show the men. The the next Saturday he left for town with thirty dollars in his pocket, fifty cents from sixty men, to pay the balance on the watch. He was to return with the watch Christmas night, but of course he didn't show up, and it was his belated arrival the men were awaiting when the loggers started into camp looking for jobs.

So Fred told Louis he would go to town with him and see what he could find out. They found that Gabby had put $1.00 on the watch, changed the receipt to read ten dollars, and had later collected the balance from the crew — then came into town and told the storekeeper he didn't want the watch. He took gloves for his one dollar, went to the saloon, and said Fred's camp needed a hundred men, and then disappeared.

Fred wired Gabby's wife and she replied that she had divorced him, didn't know his whereabouts, and really didn't care to know.

Chapter Twenty-Five

FRED BLOWS HIS STACK

Well, as Fred said, you remembered every winter for something. The winter we had the smallpox — the winter we hired the wild Bohemians, and so forth. This winter we could put down as the winter we got "took." But unless something worse overtook us, we could still count it a darn lucky winter.

The days continued fair. Old Fred's temper exploded easily and often. He drove and cursed the men like convicts. A note of tension was in the air. Louis tried to talk to Fred to warn him that he was making the crew nervous and defeating his own ends. Accidents were far more likely when he constantly yelled at the men. Louis talked to the men, too. "Try to ignore him," he said. But Fred was a hard man to ignore. As one man put it, "I never ignored any man that called me a son-of-a-bitch."

One day they brought in a skilled canthook man with a broken leg. He knew his business; he was quick and sure-footed as a cat, but Fred accused him of "damned carelessness." But all agreed it wasn't carelessness: "Nothin' but Fred's infernal drivin', drivin' us all crazy."

Fred picked a younger man from Iowa to replace the injured canthook man. He was reluctant to take the job. It was he who had said he'd never ignored anyone that called him a son-of-a-bitch. Louis warned Fred: "Lay off that boy; he won't take it." Fred tried to stay away from the loading area. He rode the two old road-monkeys he had put on. If he found a bare spot in the road, they caught particular hell. They could hardly drag themselves in at night. After Fred had stayed away from his new canthook man for a couple days, he had to go around and see how he was doing. The loads weren't moving as fast as they had under Henry's canthook.

Fred walked up quietly and watched. The lad wasn't doing so good; not nearly as good as old Henry. Fred forgot that few men were as good as Henry. He was one of the best. No one had noticed Fred standing nearby. They were trying to be careful — mighty careful. It was a dangerous job, not only for the man with the canthook but for his assistants as well. They knew better than to pile the logs as high as Henry had, so they had agreed to put on smaller loads. When old Fred saw the chains go around the load he blew up. "You yellow son-of-a-bitch!" he bellowed, "you not only send 'em slower, you send 'em smaller, too."

The boy from Iowa looked surprised just for a split second. The cold fire of his eyes and the cold steel of a double-bitted axe flew toward Fred's grizzled head at the same instant. No man ever could be sure just why Fred's head wasn't split on the spot. Must have been his heavy cap deflected the direct course of the axe enough so that its razor-sharp edge just took the skin off the side of Fred's face.

The Iowa boy jumped off the load and told Fred plenty, in language as picturesque as Fred's own. Fred undid the chains and climbed on the load, ordering his men to put up four more logs, but they just stood there and looked at him. The young Iowa boy said, "Get down off there. I want my time." A chorus of "So do I's" followed.

Fred knew he had gone too far. He got down and drove the team in himself. Twenty men drew their pay and left right after dinner, and the rest of the crew would have taken off with little urging. Louis had no sympathy for Fred. He said nothing to anyone but went to his shop and kept busy sharpening saw and axes, mending broken harness and double-trees.

Fred sulked lonely in the office. That night ten more men asked for their time; all the old road monkeys; night and day shift both, so the road would go to pot fast.

At supper that night Louis sat beside the old man, but he didn't seem to know he was there.

"He's got only himself to blame. Let him work it out," Louis told me, as I stood in the kitchen area.

But the next morning the old man wanted Louis to go into town to scout up a crew. Louis refused. "Go yourself," he said. "They got to work for you, not me. You know you won't find anything but floaters now."

So Louis drove Fred into town. It wasn't easy to get men. The men who had walked to the camp and then back, unfed, hadn't helped old Fred's reputation any.

Chapter Twenty-Six

NOBODY MISSED BALDY AT ALL

For the rest of the season Fred had to contend with riff-raff, the scum of the logging camps. He was silent and sullen, but at least silent. Louis said he was sick unto his soul. He had told Fred's wife that the old man nearly got murdered, and she wrote Fred that if he didn't hold that temper, she'd leave him home with the kids next winter and run the camp herself. Louis told me, "She could do it, too. Anyone who could manage Fred could run a logging camp."

Fred tried to make the best of a bad bargain — to bear with the shortcomings, and to reward industry and skill wherever he saw it, with a bag of tobacco at day's end, or a dollar or two off on some purchase at the camp store.

But where his cook was concerned he didn't trust anyone and wasn't bashful about letting them know it. On a sunny Sunday, if Bob and I chose to walk down a road in the woods, Fred took up his rifle and sat in his office doorway. If a man came out of the bunkhouse and started into the woods, Fred brought him up short: "Come on back. The cook's out there someplace." I told Fred that maybe some of them might get mad at being treated that way. "Can't help it," he said. "You don't know who these men might be — murderers just out of the pen, even; all sorts of criminals in these floating camp crews. Can't take any chances." At heart, I knew Fred was a pretty good man.

But you could well believe Fred's warning as you looked into the dark inscrutable faces that now made up at least half our crew. The old air of general camaraderie was replaced by another feeling of guarded tension. I felt more at ease those days when little Bob was near me in the kitchen or within sight in the camp circle, or at least when I knew who he was with. However, it would have taken a lot to get Fred to fire even one of these renegades now. The balance between the original crew and these unsavory characters was about even, so in the event of a fracas we might well find ourselves at their mercy. However, the old crew excelled in the little tricks that were designed to make a man miserable and ridiculous until he was glad to move on to other parts. Fred did not attempt to stop this horseplay. He only hoped the old crew would realize how much we needed even the worst of these men, and be tolerant. The men were not concerned about a man's past. If he behaved, no questions were asked; but when a loud-mouthed character like Old Baldy, with a misplaced funnybone, came in, things began to happen.

Seemed like old Baldy had been vaccinated with a phonograph needle. He never shut up, day or night, until someone threw a boot at him or threatened to slit his throat.

Jim was cutting meat when Old Baldy happened along. He kept pushing Jim's elbow, not once but several times. No one enjoys that stuff when he's holding a sharp knife in his hand. Jim would start to cut the meat and Baldy would say, "Here, Jim, this is the way," and he'd grab Jim's elbow and push his arm back and forth. Jim was a good-natured guy, but after several admonitions to "cut it out" and "get away," Jim laid down his knife and turned slowly. Old Baldy opened his big mouth to laugh. Jim laughed, too, but his brown hand shot out and got a good hold on Baldy's whiskers. He led Baldy into the bunkhouse and laid his face on the grindstone in the corner. He put his foot on the pedal. Around went the coarse stone. It ground off half of Baldy's beard. Then Jim let him go and went back to his own work, Baldy was laughed at all evening.

Several days later Jim was standing on a bench to get some canned stuff off a top shelf. Just as he stood balancing himself to step down with a gallon of tomatoes in each hand, Baldy came along and very gently rocked the bench back and forth. Jim could easily have been badly hurt, but got off the bench unhurt. Baldy was moving fast for other parts, but Jim overtook him and dragged him into the bunkhouse again, opened the stove door, and held the side of Baldy's face there until he singed the whiskers down to the skin. This sobered Baldy a good deal. He was quiet all during supper, but after eating the men organized a game. Baldy was "it," and every time he didn't answer right he got a kick in the pants. I knew what was up. Baldy could either give up or shut up. The next morning Baldy was a sad sight. He could hardly move. I know he must have been black and blue. He started out as if to go to work, but stepped out of line into the shadows near the office and asked Fred for his time. Nobody missed Baldy at all.

Chapter Twenty-Seven

OLD TOM NEEDED KILLING

Sundays seemed to bring a few visitors to our camp. Other camp bosses of the region came to discuss possible campsites for the next winter, or a timber cruiser on his rounds would put up over Sunday. But one day, just as Jim and Bob and I were sitting down to our meal, in popped our farm neighbor, Mrs. Madden with her husband and the seven Madden daughters. We invited them to fall to and share our food and they sure enjoyed the meal. Mrs. Madden said it saved her a big chore not having to get Sunday dinner when she got home. Feeding her brood meant no more to our supplies than if a sparrow had taken a peck from a bushel of corn. She had never been inside a lumber camp before, although camps had been around her neighborhood for years. She had met me the day I came to camp when Fred left the car at her place. She said she felt different about coming in for a camp visit than she would if there were only men.

After dinner, with the dishes done and the table set, I showed her around. I pointed out the bunkhouse and she wanted to go in for a look. She had never seen a bunkhouse before, she said. Well, neither had I, on a Sunday with the place full of men. I hesitated, but she insisted that we knock and ask if we might come in. So I knocked. "Would you men mind if this lady came in? She has never seen a bunkhouse."

"Sure," they chorused. "Come right in."

So in we trooped — big, fat, jolly Mrs. Madden and I and the seven young Madden girls. Madden had stayed in the cookhouse talking with Fred and Jim. Mrs. Madden took a good look around. She was very interested. She saw the wire frame around the stove, full of drying sox, and then she realized something. "Phew!" she said, "it stinks in here!"

"Well, what do you expect, lady?" came a big voice. "We have beans every day and wet socks to dry every night."

Mrs. Madden seemed ready to go, and as she moved toward the door the back of a hand flashed out and smacked her ample back-side. Mrs. Madden turned, fire in her outraged Irish eyes.

"Hold your water, ma'am; hold your water," said the bearded offender.

Several men moved toward the bold one. "Here, you apologize to the lady."

He rose to his feet and made a gallant bow. "Ma'am," he said, "I told you to hold your water. Well, you can let 'er flicker now."

Mrs. Madden slammed the door in a rage, but after I explained that the man really meant he was sorry, and that, after all, she was invading a world where no woman should be, she agreed she should have taken my advice and stayed out of that bunkhouse. Fred said he was relieved that so many females got in and out from among that gang of cutthroats with nothing more serious happening.

Later that week as we were about through with our noon meal preparations, a man came into the cook-house. He asked for a meal, which I prepared for him. He was an intelligent, clean sort of fellow. I hoped he was looking for a job. Fred would sure be glad to get such a fine appearing man. He said he was lookng for work and he'd look the timber over and see Fred at suppertime. He walked slowly south, seeming to be taking in our situation. I was hoping our setup pleased him. "He seems like a nice fellow," I said to Jim. "He sure does," Jim agreed. A little later Fred came along and we asked him if he'd seen the man. He said "No." While we stood talking, the Sheriff and a couple of men came in, and said they were looking for a fugitive, and described the man we were just talking about. They said he had killed Tom Kelly, a bawdyhouse keeper in town. It seemed that he had come clear from Hungary to kill Kelly. Kelly had ruined his younger sister, they said. Well, the Sheriff and Fred talked on about

Kelly, how he had needed killing for a long time. Funny a fellow had to come so far to do the job.

"Which way you say he went, Fred?"

"South, Mrs. Jimmie said."

"Well, I'll be getting on," the Sheriff said, heading down the north road.

"Yep, old Tom needed killin' a long time," Fred remarked, as he watched the Sheriff disappear around the first bend in the road.

"He's going in the wrong direction, Mr. Fred," I said.

Fred scratched his beard and grinned. "You've heard about tempering justice with mercy, haven't you? Well, up here we temper it with common sense."

And so as far as I know, that sheriff is still going north while the killer may be still headed south. Who killed Tom Kelly was never put on the books; he washed his hands in our bunkhouse, he ate at our table, and he went his way.

Life went along now at a pretty even pace. Little things happened to break the monotony, but we had no serious trouble. Things were shaping up better than Fred had a right to expect after he'd spited himself with the men.

Old Fred was a little worried one morning when four men awaited him at the office and asked for their time.

"If you were going to quit, why didn't you quit Saturday? Why wait till Monday? Anything wrong with this camp?" Fred asked as he handed out the checks.

One man answered, "Aw, we're not going to stay where we get burnt meat for breakfast."

"Burnt meat? I never saw any burnt meat here. Mrs. Jimmie wouldn't burn it in the first place, and if she did, she'd never put it on the table."

"Well, we had it at our table," they all agreed — and went down the road.

Fred came to me, looking puzzled.

"Mrs. Jimmie, did you burn some meat this morning?"

"No, I didn't burn any meat," I told him.

"That's sure funny. I didn't think you did, but four men just quit. They said you gave them burned meat for breakfast."

"Gosh, they must be crazy."

"That's what I thought, too, but they seemed so sure about it." Fred took off his cap and scratched his head, puzzled. The door opened and our driver came in. He sat at the smaller table where the men had sat who claimed they had burned meat. Fred asked him about it. "You set at the small table, Ed. Did you have burned meat this morning?"

"No sir. We had liver. Good, too."

Then Fred told him why he asked, and Ed laughed and said, "Everyone wondered what was the matter with those guys. They refused the meat every time it was passed, and looked at the rest of us so funny while we ate. I suppose they wondered why we were eating burnt meat without a gripe."

"That's a good one," Fred said. "Guess they were too dumb to be worth much to me anyway. Next time you have any burnt meat like that, Mrs. Jimmie, I hope our table gets some."

Chapter Twenty-Eight

DOC STILL HASN'T PAID HIS TAXES

I guess if I had a favorite among the men it was little Shorty Newberry. He was seventy-five years old, but it was hard to believe, he was so sturdy and tough. He looked and acted like a good-tempered grizzly bear. That is, his appearance reminded you of a grizzly but he wasn't vicious at all, although he was so strong no man could escape him once he got his powerful arms around him. Old as he was, his favorite sport was wrestling, and on Sunday he loved to take his position sitting on a stump near the bunkhouse door, challenging all comers to knock him off the stump. Very few knew his age, and attacked him with all their strength. And while there were times I feared the old warrior was about to meet his match, somehow they never quite succeeded.

It was Shorty's job to haul in the grub once a week and, of course, bring the mail along, too. He usually got in just before dark when I was rather busy in the kitchen, and he'd make a dash in with the mail before taking the team to the barn. He wanted me to sort the mail and put it around at the men's places. In the first place, I knew all faces but few names, and in the second place I didn't have time. Shorty made various excuses for his not taking care of delivery, such as "late," "in a hurry," "forgot my glasses," "got something in my eye," and so on. He'd come in late and sit with the paper in front of him.

"Awful wreck here, Mrs. Jimmie," he'd say. I'd look. "You got the paper upside down, Shorty." Shorty would laugh and turn the paper. Then one day he said he didn't see how they could sell a set of harness for $6.25. Sounded awful cheap to me, too, but when I looked it was $62.50.

Then one day I knew there was something wrong with Shorty. Just what, I didn't know. He always seemed a well-balanced little guy; no bats in his belfry; but he showed me a cross-word puzzle in the paper. "See," he says, "these black squares are the people that didn't pay their taxes, and the white ones are people that did. I paid mine, but here's Doc Jones, across the road from me; he didn't pay his taxes yet."

For several weeks Shorty looked to see if Doc Jones had paid his taxes yet, but the square remained black.

"See, Doc still hasn't paid his taxes yet and he's rich. He's a big shot. But I paid mine. I paid mine a long time ago."

Then one day when Shorty brought the mail in, there was a letter for him. When I handed it to him he asked me to read it to him. I hesitated. "Shorty, you better get your glasses and read it yourself. This is personal — from your wife to you." Then Shorty told me. The arrogance and pride that was Shorty dropped away. I looked into ashamed little-boy eyes, suffering eyes, as he said, "Mrs. Jimmie, I can't read or write. I never learned."

Poor little Shorty! No wonder I had thought him queer when he tried to read the paper. I was glad I had kept still when he told me the crossword puzzle was a tax report. I didn't want to hurt him, and he had such fun gloating over Doc Jones' nonpayment of taxes.

But now I opened his letter. It was very brief. It was the only letter little Shorty had gotten since camp began. There was no "love" or "kisses." Just "Send the money for the taxes. They got to be paid."

Poor Shorty! I was glad we were alone. It was an awful hurt. He stumbled out and stayed away from me for two days. Then he came in with the paper again, and started to tell one of the new men about the tax report. Before I could warn the new man, he gave Shorty a sharp look and said, "Are you nuts? That's no tax report. That's a crossword puzzle." And Shorty slunk off again. I sought him out the next day.

"Come on, Shorty. Don't feel so bad. But I want to know just how come a smart little guy like you never learned to read or write.

I know your brother Danny, and he's one of the village fathers. He sure can read and write, can't he?"

"Yes, he can," Shorty said. "We didn't have a chance to go to school till we were about twelve. Danny went to school, but I ran away from home and I been on my own ever since."

"Well, Shorty, you got along for seventy-five years without reading or writing. You have a wife and children — and friends. You've got a nice farm — so why worry about this now? I like you just the same. Just come back in the kitchen and visit and have fun, and don't feel bad. Lots of men can read and write and even went to college, but they haven't got half what you have, and never will have."

The next day Shorty was back, his old usual self. But he contented himself with whittling now, scorning any notice of the tax report.

Chapter Twenty-Nine

WHAT IN HELL IS THAT STINK?

Old Fred had been mighty pleased with my work and had often told me I had disproved the old tradition that a woman in camp brought only trouble. He said I had fed the men better than they were ever fed. Better and at a great saving. He'd be proud to hand me the promised bonus. That was why, when I made one gosh-awful blunder, I sure didn't want Fred to know. I sure would lose some of his good opinion — maybe the bonus, too. Or get fired in disgrace, even.

This is what I did: I wanted to get the big meat kettles off the stove earlier one day, so I saw no reason why I shouldn't put the meat on to cook the night before, and then when I got up the meat would be cooked and ready to set off the stove. Very clever plan, Jim agreed.

We set the big kettles on, full of pork — half a hog in each of the two big kettles. Jim was to get up a couple of times during the night to put more wood on the fire. This he did not do. We both slept right through until 5:00 A.M. When we looked in the meat kettles, three inches of foam appeared on the top and a strange sour smell greeted our nostrils. I did not know what had caused this strange occurrence, but it was very evident that the day's meat was spoiled. I saw Louis pass the window, and motioned him to come in. We did not discuss the cause of our problem just then. The immediate need was to dispose of the mess without old Fred knowing it. Louis and Jim dumped the two kettles in the sink to send the liquid down the drain. Then Louis put some dry tamarack in the heating stove and we piled the meat on top, covered with more precious tamarack. Saying silent prayers, we went outside to note the wind was taking the smoke toward the west and away from the campsite. With a half hour to go before Fred would emerge from his office, we breathed easier. But five minutes before Fred would come out of the office door the wind shifted right over the cookhouse, and when Fred opened his door the smoke hit him square in the face. He made a face and bellowed, "What in hell is that stink?"

Of course, no one could enlighten him, or at least the three who knew didn't make any effort to answer him. The men ate breakfast. The stove, full of meat, wasn't very far from old Fred's back.

"Sure got a hot enough fire in here," he said. "Don't burn all that dry tamarack up at once. Damned scarce stuff."

I was afraid the extra hot fire might cause him to open the stove door and look in. Old Fred finished breakfast and went out, looked at the smoke, and sniffed the air. "I know what that is," he said. "You see the wind has shifted. That smell is from the glue factory in Winter. Boys, that means Spring ain't too far away. When the air carries like that, we got to hit the ball."

Louis came in. "Wasn't that something? He thought it was the glue factory in Winter. Put this down as your lucky day."

"It's at moments like these that I feel close to God, Louis," I said, "I've been praying all morning."

Well, we had beef for dinner, and only God and we three ever knew that the smell had not come from the glue factory.

Chapter Thirty

SPRING BREAK-UP

But the changing smoke had meant Spring was in the offing.

It seemed no time at all until the family grew smaller; then at last our final meal together. The teams and their drivers lined up to leave, and were waiting to say goodbye. Jim was driving Maud and Bess. Bess was nervous, as always. Maud's expression seemed to say that she didn't give a damn one way or the other.

But for all of us there were regrets — regrets that the winter was over, because many of us would never see each other again. I threw my arms around Shorty and gave him a big loud kiss: "Goodbye, you old Irish rascal," I said.

"Mrs. Jimmie," he said, "I hope you never forget the winter you went a-loggin'."

"I won't, Shorty."

Little Bob wanted to go with Jim in the sleigh, but it would have been too long a trip. We were going later in the car with Louis. Old Shep was more than ready to take her place with Jim and the horses, homeward bound. Pussy had taken up her abode in the barn with Rosalee, the camp's little Jersey — warm milk twice a day, and a warm bed any time she chose. When we went in to milk, we'd find Rosalee lying down and Pussy curled up on her back, purring with contentment. I had no assurance I could offer Pussy such comfortable arrangements back home. Rosalee's new owner thought it a shame to break up such a beautiful friendship, and was happy to have it continue in his barn, so Bob and I said goodbye to Pussy.

A feathery skiff of snow had fallen in the night. It would speed the trip home if it held through the day. All the sleighs had a passenger or so, glad to hitch a ride to some point or other on the way home.

We waved and called back and forth until they were out of sight.

"We'll come and visit you this summer."

"Let's all go a-loggin' next winter."

"Take good care of little Bob."

George and young Ed, and Shorty, and Jed Barton and his boys, all faded from view.

Louis and I turned to to help Fred with the last-minute packing of equipment. After the noon meal, Fred told Louis he'd better be on his way, taking Bob and me home so we could have the house warm and supper cooking by the time Jim pulled in with the team. We put all our belongings in the car; Louis wrapped us in a big blanket and put warm stones at our feet, and we were on our way.

I was quiet, my thoughts on the good days I was leaving — plenty of good food, warm, cozy house, work to do, people in and out, jokes and fun.

"You look awful glum for a girl that just got herself a hundred bucks bonus," Louis said.

"That was wonderful of Fred," I said. "That's the only part I'm happy about — that check for three hundred and fifty dollars in my pocket. It means Bob can go to the hospital right away, but I'm going to miss everything else. This money looks awful big, but Louis, it won't last forever, and I don't believe I can take being hungry and cold again, ever. And I want little Bob to have everythng he needs — including an education. He can't get it up here."

"Well, you made a name for yourself. Last fall everyone thought you were crazy; now they think you're a darn smart girl. You can cash in on that. Fred says you can cook for him as long as he has a camp, and you won't have to ask him. He's going to ask you."

"That's all nice, Louis, but my idea of being a wife is staying home taking care of little Bob and his little brothers and sisters. I wouldn't want to go to camp every winter, I guess."

"Yes, I know what you mean. The country has gone to pot. I don't blame you for being restless."

"Look: Now we're about ten miles from home. Notice the gaunt, empty houses, one after another — forlorn ghosts on every quarter-section."

We stopped at the general store to gather the winter's gossip.

"The Charley Walsh's? Old Charley got took bad; died in the Vet's Hospital in Milwaukee. Charley always said old soldiers never die, but it seems they do. Charley's timid little widow made her way back to kinfolk in Iowa." And the eyes of the Walsh home stared and saw nothing.

"Withered old man Slacksaw went out in the barn one morning; found his only horse dead. He took the halter rope off and hung himself. Old lady Slacksaw sucked in her breath through toothless gums and tried to barter her repulsive charms for food, fuel, or anything to keep life in her. A skinny old hag. Even this profession was unprofitable here."

Mrs. Staden went insane after twins were born. Some wives had run away — left their men; and some men had left their families.

I told Louis I was ready to go. I had bought a few things. Needed more, but I couldn't stand more of this kind of news.

We knew most of the Onesti story. Papa Angelo and Mama Rosa came to this country with ten strong bambinos; they started on 160 acres, they worked hard, but "no can make money," Angelo say, and JC say must have money. JC moved them from place to place, each place smaller and not as good as the other. Each year one or two bambino leave; too hard for Papa Angelo to feed so many. At last all are gone, to work here or there. The dream of "Onesti and Sons" had gone with them.

Then JC moved Papa Angelo and Mama Rosa to ten acres of swamp land near an old abandoned logging camp. "Good marsh hay here, Angelo. Keep your animals fat. Blueberries, too, for you and Mama Rosa." JC pounded Papa Angelo on his broad back, smiled his big genial smile. "You'll be happy here and now you don't owe anything. You'll never have to move again."

Angelo turned away from the tears in Mama Rosa's dark eyes. A smoldering hatred in his own followed JC's car down the old logging camp, where he lingered. What was he poking around there for? Why didn't he get out of sight?

"I hate his stinking guts! The marsh grass keep my animals fat? Cut their bellies to ribbons, it will! Blueberries — about once in five years, if we beat the fires or the frost. No, we don't owe anything. Never have to move again — just sit right here till we die."

Then Angelo got an idea. JC was still at the old campsite. Angelo started on the run. Hearing him approaching, JC turned in surprise and anger.

"JC, excuse please, I frighten you. I just think why cannot Mama Rosa and me live in camp buildings here? Nobody use them for years; we take care very good! High here — Mama Rosa would be happy; in swamp, sad — Mama Rosa cry." Angelo was sure JC would like this plan — life and warmth in the old logging camp. But JC shook with rage. "Angelo, I don't want you or anyone else on this campsite. I'd better not see any signs of your being around."

Angelo had never seen this JC before, nor — please God — let him never see that face again — no warm smile on the large mouth, no friendly light in the blue eyes. The mouth was a thin straight line. The blue eyes were cold steel. Hate burned there — hate for all the little people; hate for their whimpering, crying voices — most of all hate for Angelo himself.

Angelo ran back to the swamp and to Rosa. "Mama mia, I make cabin warm for you. You see — we be happy here. Children come sometimes and visit us."

Angelo held her hand warmly. "Mama Mia, Mama Mia." He did not tell her what had passed between himself and JC. Better only to talk of happy things.

And so Angelo prepared for the coming winter. Each day Mama Rosa looked more sad and her step was heavy and slow, her shoulders stooped as if she were always tired, though Angelo let her do no heavy work. The swamp mist seemed to sink into her bones, and she coughed a good deal. Angelo tried to make jokes, and he made her drink a little good wine each night from the bottle their son Aldo had sent. Even after the wine Rosa did not smile easily, and her sleep was troubled and her cough grew worse.

Angelo worried about Rosa, and wrote to the children. Those who could sent a few dollars.

Vito wrote, "Come to the city, Papa Angelo. We take care of you and Mama Rosa." But Angelo said, "No, I stay; if only you can take Mama Rosa where she be warm and dry."

But Mama Rosa wouldn't go to city. "Children need their money; I not be a burden on them. We both stay right here, Angelo; I be fine; just little cold."

But when Vito came in the spring he found Mama Rosa dead. She had been dead for some days.

"I can't leave her, Vito. So I stay, I wait. Nobody came, I guess everybody dead; country dead. We see no one all winter."

Vito buried Mama Rose, "near so I can keep her grave nice," Angelo said.

"Let's take her to the city, Papa Mia, and you come there too," Vito urged his father. Angelo would have none of that, and when Vito suggested "a small spot near the old logging camp, close to you, but out of the swamp," Angelo roared, "No, no! In the name of God, no! That's an evil place! I will stay in the swamp. Leave Mama Rosa here with me."

So it was. Rosa was buried under the sharp marsh grasses.

Each year several of the children would come to visit Angelo and try to coax him to the city. Now there were plump grandchildren, and they brought a smile to the old man's face, but he wouldn't leave.

Vito came in the fall to try once again. It worried all of them to think of Papa Angelo out there alone through another northern winter, but he said he wasn't alone. "Rosa is here, you know."

"But Papa, you'll starve here — no game any more."

"I see tracks, Vito."

"What kind of tracks, Papa?"

Angelo did not seem to hear the question. He seemed lost in deep thought. "Tracks," he said, "tracks by the old logging camp."

So Vito wandered over there the next day. He saw no tracks; that is, not animal tracks, only man tracks — Papa Angelo's own, Vito decided.

"I saw no tracks around the logging camp but yours, Papa Angelo."

"Not my track, Vito, I never stop there. JC look like devil, tell me to stay away. I never go there."

"Well, then, how did you see tracks there if you never go there — and what kind of tracks?"

"I just sit on edge of my swamp, look on campsite, see tracks — plenty tracks."

And sure enough he did. Vito watched him. Angelo sat concealed and watched the campsite, never an inch on JC's ground. He sat there from early evening till late, sometimes. Vito never called; he only watched and waited and wondered — had Papa Angelo lost his reason? And what could he, Vito, do about it?

Then one night before sunset Angelo went again, sat in the same place and waited. He had his deer gun; he always had his deer gun; must be a deer he expected to see.

He hadn't been gone twenty minutes when a shot shattered the evening stillness. Vito made the distance in great leaps, fear in his heart, but Angelo stood proud. "A buck, Vito, four points, I think." Angelo pointed, but did not move forward. Vito walked about ten yards and saw the prone figure of a man — a big man who usually wore a big genial smile on his face. Now he had no smile and almost no face.

Vito turned and looked at Angelo. He knew Angelo knew it was no buck deer he shot; he knew the tracks were JC's, and he knew Angelo killed JC, and knew it was JC before he pulled the trigger.

"It was an accident, Vito," Angelo said, very calm now. "He had his arms up and the rifle in his hands, pushing the brush away, and I thought it was antlers."

"Yes, it was an accident, Papa; but the bullet you used was no accident, was it?"

"No, I make that special. I know a rifle bullet wouldn't do that to a man's face."

So Vito made tracks for the village store to call the sheriff.

The sheriff asked no unnecessary questions since the shooting was an accident. Angelo, having seen no human for months except his own son, could not under the circumstances be blamed for taking two arms and a rifle waving through the brush at sundown, for a brace of antlers. If he had no money for the proper ammunition and had concocted a bullet of his own from odds and ends, that, too, was more or less an accident. And JC, who had not been seen in these parts for these past two years, happened to appear by the old logging camp — that, also was an accident or an act of God.

Louis and I left the store and took Vito back to Papa Angelo on our way home.

After the sheriff took over and the verdict of "accident" was handed out, Papa Angelo announced he was ready to take Rosa and go with Vito to the city, where he could see his children and their bambinos every day.

There was much speculation concerning the accidental shooting, around the village store when the scattered trappers and idlers drifted in.

"What in tarnation was JC doing in the country secret-like, and hanging 'round the old logging camp?"

"Bet he had money buried there."

Some tried to question Vito and Angelo before they left. Vito said he knew nothing except what happened after he heard the shot. Angelo muttered strangely at questions.

"He's crazy," some said, "Maybe like a fox."

Chapter Thirty-One

MY MARLIN 25-20

We got to our home place about 3:00 o'clock in the afternoon. Louis got both stoves going, then he had to go back. He and Fred had a couple days work at camp yet. He started to unload our stuff and bring it in.

"Hey," I said, "you're bringing in your stuff, too. Here's a box and your rifle."

"Well, the box is yours. Wait, I'll show you," and Louis opened the box to reveal a beautiful set of carved shelves, filled with fans and cupids, animals, and a tiny replica of our campsite, with a little wooden Bob driving a wooden team to the wooden barn, and the cookhouse with a small wooden me in the door. I took everything in with amazement and delight.

"Louis, I never hope to see anything like this again in my whole life. You're an artist!"

"Well, now, I didn't do this all alone. I had the idea, but nearly every man in camp helped. It's our memento to you — so, as Shorty said, 'She'll remember she went a-loggin'. The men were too bashful to present it. They wanted me to give it to you when you got home."

As Louis buttoned his coat and pulled on his big mitts, I picked up his rifle to hand to him. Most men carried 30-30's or even bigger, but Louis had bought this beautiful Marlin 25-20 at the beginning of camp. Everyone kidded him about his lady's gun. I admired it every time I went into the office, light and beautiful, an ivory bullseye set into the polished stock; ivory bead on the sight. "There's a gun," I'd say, as I sighted, and held it with longing fingers.

"Fred bought this gun from me this morning," Louis said.

"Gee, I'm surprised at that," I said. "I though he always laughed at it."

"I was surprised, too. But you're going to be more surprised when I tell you he bought it and then told me to give it to you. Fred said you worked so hard all winter, and he knew you wouldn't spend a cent of that money on yourself. Said you'd been in love with that gun all winter and, as for him, he was sick of looking at it."

"Louis, I'm so happy I don't know what to say. I'm just afraid I'm going to bust something!"

"Well, I got to go before that happens."

Chapter Thirty-Two

I'M GOING WHERE YOU GO

That night Jim and I lay in bed and talked till way late. It seemed sort of good to know you didn't have to get up at 5:00 A.M. Jim was glad the winter was over; being inside all day long had not been easy for him.

We agreed that we had better get Bob to the hospital and have that job over and be back home again before spring work got underway.

As we talked we seemed aware of movement about us — between the walls, between the floors — the darned house was alive! Jim said something had sure taken over. We'd investigate tomorrow. But soon we identified our house guests. We attempted to ignore the commotion they were making and go to sleep. But something seemed to be bouncing through the springs of the bed. A little body walked over my arm; it took little nibbles; it gave a squeak of delight. I struck out with my other arm. We both sat up. "It's mice," we said, together. "Dozens. Maybe hundreds."

Jim lit the lamp, fixed a trap, and got into bed. "Snap!" Before he even got settled! He removed the dead mouse, set the trap, came back and started to crawl into bed. "Snap" again! Over and over we set and reset that trap. "They're getting wise. Won't be more so quick" Jim said. But it kept on until Jim had caught twenty-five mice. It was 12:00 o'clock. Jim set the trap for number 26, and said, "To hell with it! I'm going to sleep."

The next day we bought more traps and really went to work. I had a box upstairs, about two feet square, full of pieces of cloth for making patchwork quilts. This was a mouse nursery; a wiggling mass of tiny, pink baby mice. Lord, it made me sick!

We left the next day for Chippewa Falls to arrange for Bob's operation. As soon as it was over, Jim left for home, while Bob and I stayed on two weeks until he could walk again.

When we got home very little remained of the check I had brought from camp. It paid the doctor's bill and the hospital bill. In addition, I had stayed right by little Bob for the two weeks and had a board bill of my own, train fare, taxi fare, a few clothes and things we needed for the house. Nothing frivolous.

Jim's check, with Maud and Bess's wages added, had been as big as mine, but I came home to find that it, too, had shrunk. Jim had bought staple groceries, horse feed, and seed for planting. He had had the horses fresh shod. Jim also made the trip to the county seat and paid those outrageous taxes.

He had brought our little Jersey cow home and she had thrown her calf four months ahead of time. The vet said she was only fit to sell to the butcher. We had been stung on the deal. The vet said he had told the previous owner she would never bear a calf.

"The only thing that seems to produce around here," I said, "are those darned rocks. Look at them, Jim. Three for every one we had last year."

Jim knew I had got the bit in my teeth since camp. I was restless. I tried to help pick rock and get the garden ready to plant, but it seemed my back ached sooner and stronger than it ever had before. I stopped more often and sat and stared. I didn't care if I ever picked another rock. But little Bob felt good again these days, with the old truss gone. He sure got over the ground. Jim said he'd get the stuff planted and then take off for the canning plant again. He didn't want to quit. Everything would work out for us.

One mellow, moonlit night the smell of spring was so rich and wonderful, we had left all the windows and doors wide open. We had just slipped off to sleep when it seemed, as if in a dream, that there was a soft nicker, and then another, pleading and urgent. I shook Jim's shoulder. He listened. It came again — almost like a human cry.

"It's Bess," Jim said. "She's going to have her colt."

"Can I come too?" I asked, as I put a coat over my night clothes.

"Sure. But be quiet and don't frighten her."

So we went to the pasture gate. I stayed a little distance behind. It was Jim, her master, she called. She came and nuzzled him gently and led him out away from the wire fence. She trembled and her body was wet with nervous perspiration. It seemed she wanted Jim to tell her what to do. She didn't know whether to lie down or not; motherhood was new to her.

"Lay down, girl," Jim said, and stroked her suffering face. She lay down with a great sigh, exhausted by sheer fear. Jim and I stepped back a bit. She struggled to rise, frantic eyes pleading. Jim hurried to reassure her: "It's all right, girl. Just take it easy." Her beautiful body heaved and thrashed. Deep groans rent the air, ending in a piercing scream that I'll never forget. Jim, now in close attendance, pulled away from her kicking hooves what looked like a light-colored transparent bag. I ran to Bess's head. I sat down and put her head on my lap. I loved her and patted her while big tears splashed down her face.

I thought something was wrong; I could see no colt in this strange mass, until Jim cut the cord between the mass and Bess and then with his knife opened the bag. There, struggling to stand, was the most beautiful little colt I ever hoped to see.

Bess arose so fast she knocked me over. She cleaned her baby and loved it gently. In less than ten minutes the new colt was having its first meal, its little tail going round and round like a windmill.

I was so excited I couldn't sleep all night. I'd get up every hour or so to see how things were going.

Bess was a good mother, and so proud — and well she might be, for her daughter, Dolly, was a beauty.

Old Maud was mighty envious, but that old reprobate could have had one of her own if she weren't so ornery. She acted coy and receptive to her prospective spouse until the final moment, and then she kicked the very devil out of him, and no stud owner would attempt to breed her after that.

Jim had a standing offer for little Dolly as soon as she was weaned. He said we'd better let her go. She would need a lot of feed the first year and it would be expensive. He didn't want to take a chance on her having to miss some meals and not develop into the fine creature she had every right to be.

Jim going away to work again! Little Dolly sold! Just Bob and I alone in miles of forsaken country! I couldn't stand that again. I told

Jim I was going to take Bob and go and stay with friends near Chicago while he was gone. Get a job.

A few days later when Jim was on the back forty, some horse buyers came along — wanted to look at Maud and Bess and little Dolly.

"Tell Jim we'll take them," they said.

I laughed. "Jim wouldn't sell them. Just Dolly."

But when Jim came in and I told him about it, he said, "It's O.K. by me. It's a deal."

"What are you going to do?"

"I'm going where you go," he said.

— THE END —